The Institution of Electrical Engineers

UK MANUFACTURING

FACING INTERNATIONAL CHANGE

This document was prepared by Professor J C Levy under contract to the IEE.

The Institution of Electrical Engineers in amalgamation with the Institution of Electronic and Radio Engineers and the Institution of Manufacturing Engineers.

Published by the Institution of Electrical Engineers, Savoy Place, London WC2R 0BL

© 1994 The Institution of Electrical Engineers.

ISBN 0 85296 615 6

The cover shows an extract from a cartogram of the distribution of world GNP. A complete diagram with country sizes in proportion to GNP and a key will be found on pages 102-103. This appears courtesy of Myriad Edition Ltd from Kidron Segal, The New State of the World Atlas. © Myriad Editions Limited, 1994

Table of Contents

EXECUTIVE SUMMARY vii

International Changes vii

Changes in the UK viii

Conclusions-Challenges and Responses x

INTRODUCTION 1

1. THE UK MANUFACTURING SERIES 1

1.1 The Range of the Report 1

1.2 "Manufacturing" 2

PART 1 3

INTERNATIONAL 3

2. THE EUROPEAN COMMUNITY (EC)/
EUROPEAN UNION (EU) 3

2.1 General 3

2.2 Germany 5

2.3 Germany and France 8

2.4 EU Research 9

2.5 The Enlargement of the European Union 10

3. EASTERN EUROPE 11

3.1 The unsettled scenario 11

3.2 An example in telecommunications 12

3.3 The Way Forward 14

4. THE UNITED STATES 14

4.1 General Survey 14

4.2 The North American Free Trade Agreement
(NAFTA) 17

4.3 Latin America 18

4.4 The USA/Japan Trade Dispute 18

4.5 The USA - Conclusion 20

5.	**JAPAN**	20
	5.1 General	20
	5.2 Japan and Asia	22
	5.3 Japan and China	22
6.	**THE PACIFIC RIM**	23
	6.1 A Survey	23
	6.2 South Korea	25
	6.3 China	26
	6.4 Asia-Pacific Economic Co-operation (APEC)	29
PART 2		31
CHANGES IN THE UK		31
7.	**THE POSITION OF THE UK**	31
	7.1 UK Performance (Key Indicators)	31
	7.2 Output	32
	7.3 Productivity	34
	7.4 Share of World Trade	38
	7.5 Import Penetration and Balance of Payments in Manufacturing	39
	7.6 The GATT Round	42
8.	**CHANGES IN UK DOMESTIC FACTORS**	44
	8.1 Inputs affecting UK Performance	44
	8.2 Education and Training	44
	8.2.1 Schools - National Curriculum	45
	8.2.2 Schools - A levels	46
	8.2.3 National Vocational Qualifications (NVQ)	46
	8.2.4 Training and the TECs	48
	8.2.5 Universities	49
	8.2.6 Engineering Degree Courses	50

	8.3	Research and Development	52
		8.3.1 The White Paper	52
		8.3.2 Government Research and Development	53
		8.3.3 The Research Councils and the Universities	56
		8.3.4 Industrial spending on R&D	58
	8.4	Innovation	62
		8.4.1 Innovation - does it pay?	62
		8.4.2 A German - British Comparison	63
		8.4.3 How to Innovate Successfully	65
	8.5	Management Style and Quality	67
		8.5.1 UK Shortcomings	67
		8.5.2 Concurrent Engineering	67
		8.5.3 Total Quality Management (TQM)	69
		8.5.4 How to improve	69
		8.5.5 Conclusion	72
	8.6	Size of company	72
		8.6.1 Small and Medium Sized Manufacturing Enterprises (SMMEs)	72
		8.6.2 Measures to improve the performance of SMMEs	73
	8.7	Finance, Investment and Government Policy	74
PART 3			77
CONCLUSIONS			77
9.		Challenges and Responses	77
	9.1	The Broad Picture	77
	9.2	International Challenges	77
		9.2.1 The EU	77
		9.2.2 Eastern Europe	78
		9.2.3 The USA	78
		9.2.4 Japan	78
		9.2.5 The Pacific Rim	78

	9.3	Domestic Challenges	79
		9.3.1 The 4 Key Indicators	79
		9.3.2 The 6 inputs	79
	9.4	The UK Response	80
		9.4.1 General Responses	80
		9.4.2 Area-Specific Responses	80
	9.5	A MANUFACTURING EXPANSION SCHEME: Four National Initiatives And Their Underpinning	82
10.		REFERENCES	89
11.		BIBLIOGRAPHY	91
12.		Annex A - EXECUTIVE SUMMARY OF 1992 REPORT	93
13.		Annex B - USA TECHNOLOGY SUPPORT PROGRAM	95

vi

EXECUTIVE SUMMARY

This third report in the "UK Manufacturing" series features the tremendous economic and social changes which have taken place world-wide during the past couple of years and assesses their impact on UK Manufacturing. It makes proposals - expressed as four linked initiatives - to show how the UK can respond to expand its manufacturing base and increase its share of world trade.

PART 1

International Changes

Rarely have so many international changes affecting manufacturing competitiveness occurred in so short a period of time. Part 1 of the report discusses them in some detail.

• The recession in Western Europe and the difficulties of major industries such as steel and chemicals	**section 2**
• The lack of competitiveness of the European Union vis-à-vis the USA and Japan	**section 2.1**
• The absorption of East Germany into the European Union and the consequent problems	**section 2.2**
• The forthcoming enlargement of the European Union	**section 2.5**
• The disintegration of the USSR and the new challenges	**section 3**
• The launch of the North American Free Trade Area (NAFTA) and the technology plans of the USA	**section 4**
• The possible ending of the Japanese economic miracle, the USA/Japan trade dispute and Japan's trade ambitions in Asia	**section 5**
• The development of the "Pacific Rim" countries, particularly the "little dragons" of South Korea, Taiwan, Hong Kong and Singapore	**section 6**
• The rapid expansion of the Chinese economy and the opportunities it affords	**section 6.3**
• The growing activity of the Asian Pacific Economic Co-operation (APEC) organisation	**section 6.4**
• The success of the General Agreement on Tariffs and Trade (GATT) negotiations	**section 7.6**

In total, the changes set the agenda for years to come and could add up to a period of even fiercer manufacturing competition than to date. The accent on free trade in GATT, NAFTA and APEC while welcome in itself, threatens a continuing shift of manufacturing to low wage cost countries and, at the same time, could strengthen the position of the USA and Japan as efficient, high-quality producers in a position to take advantage of the lowering of tariff barriers. The UK balance of trade problem exists mainly outside the European Community so we must take advantage of the plentiful and varied new opportunities in such different markets as Eastern Europe, China and the smaller Pacific Rim countries, not to mention the USA and Japan themselves.

PART 2

Changes in the UK

Many reports have demonstrated, that the only way to achieve a positive balance of trade is through manufacturing and it seems that the Government accepts the message. Service industries alone cannot fill the gap, even though, with manufacturing, they form an integral part of our wealth creating activity.

In Part 2 of the report the recent manufacturing performance of the UK is examined under the same headings as the original 1992 report [1]. There are some encouraging features on the restructuring of industry and on market consciousness but any upturn is relatively small compared with the need. It is concluded for the four "key indicators" that:

Output	The increase between 1975 and 1991 compares unfavourably with the USA, Germany and Japan.	**section 7**
Productivity	Despite improving recently at a greater rate than competitors, productivity still lags by 20-25% in absolute terms.	**section 7.3**
Share of World Trade	This has continued to drop and we may now be below our "fair share" for developed nations in proportion to population.	**section 7.4**
Import Penetration and Balance of Payments	The trade gap worsened in 1992 and 1993. There is little sign of a reduction in our balance of payments deficit, particularly with countries outside Europe. Although it is true that we export more per head than does Japan or the USA, we export less per head than France, Germany or the Netherlands.	**section 7.5**

The six "inputs" (so-called because they are within our own control) which can influence the key indicators now give the following picture:

Education and Training	Our population is still under educated and trained by international standards, particularly at levels below the top one. The effect of the shortfall is experienced most acutely by small and medium-sized enterprises because large enterprises have first pick of the available talent. Recent Government moves in schools and on the apprenticeship front may gradually improve this situation but considerable problems remain on NVQs, training and the funding of higher education.	**section 8.2**
Research and Development	There was actual reduction in R & D spending in 1992 and the effort is still too concentrated among a few firms. Even our best firms have, on average, a 2:1 ratio of distributed profit to R & D expenditure whereas internationally the best 200 firms reverse this ratio. The White Paper "Realising our Potential" is helpful but will need determined follow up to ensure its aims are achieved.	**section 8.3**
Innovation	Recent studies in the UK have identified the successful routes to innovation but they are not widely applied. More emphasis is needed by firms of every size and type on "incremental " innovation to complement "new product" innovation.	**section 8.4**
Management Style and Quality	Management in the UK still leaves much to be desired. Known lessons are not yet applied widely enough. Concurrent engineering and Total Quality Management are not exploited sufficiently.	**section 8.5**
Size of Companies	Our largest companies are comparatively flourishing and remain world competitive. For the small and medium-sized manufacturing enterprises (SMMEs) which employ 45% of the work force and are responsible for a third of production, most of the 50 points of improvement listed in the 1993 paper on SMMEs [2]	**section 8.6**

	remain to be assimilated. Financing problems need to be solved for SMMEs in the "growth corridor".	
Finance, Investment and Government Policy	Our plant is on average relatively old and we still lag in capital investment. The taxation system continues to penalise high-technology industries which depend upon heavy investment in R & D and equipment.	**section 8.7**

PART 3

Conclusions - Challenges and Responses

Part 3 summarises the international and domestic challenges faced by UK Manufacturing. It goes on to make proposals to improve UK output, productivity, share of world trade and balance of payments.

These proposals are expressed in the form of a Manufacturing Expansion Scheme having four linked national initiatives.

A MANUFACTURING EXPANSION SCHEME:

- A Management Initiative - People and Organisational Design

- A Technology Initiative - The Agile Factory

- A Finance for Industry Initiative - A new arrangement for Small and Medium - Sized Manufacturing Enterprises (SMMEs)

- An Innovation Initiative - Taxation of Innovative and Capital Intensive Enterprises

All these initiatives need to be underpinned by continuing effort to improve our education and training system. It is suggested that a national debate should be launched to simplify and make more accessible the complex array of qualifications and routes.

Complementary responsibility for the initiatives and their underpinning must be shared between Government, employers organisations, education and training bodies, professional institutions, individual manufacturing enterprises and the financial community.

This report is the third in the IEE series under the general title "UK Manufacturing", prepared by Professor Jack Levy.

The previous publications (available from the IEE) are :

(1)   - "A Survey of Surveys and a Compendium of Remedies", May 1992.

(2)   - "Small and Medium-Sized Manufacturing Enterprises - A Recipe for Success", July 1993.

INTRODUCTION

1. THE UK MANUFACTURING SERIES

1.1 The Range of the Report

This is the third in the "UK Manufacturing" series produced for the Institution of Electrical Engineers by Professor Jack Levy FEng. The first report "A Survey of Surveys and Compendium of Remedies"* appeared in May 1992 and the second, entitled "Small and Medium-Sized Manufacturing Enterprises"**, followed in July 1993.

In the two years since the first report the situation in the world has changed extraordinarily rapidly and in some respects profoundly.

The recession in Western Europe affecting the EU*** , the new, larger European Economic Association (EEA); the absorption of East Germany and the disintegration of the USSR; the USA economic response to the end of the cold war and the launch of the North American Free Trade Association (NAFTA); the possible end of the Japanese economic miracle, the development of the Pacific Rim countries and the rapid expansion of the Chinese economy; the growing activity of the Asian Pacific Economic Co-operation (APEC) organisation. These, and the success of the negotiations on the General Agreement on Tariffs and Trade (GATT) all add up to a substantial new challenge for UK manufacturing competitiveness.

Rarely have so many major changes taken place in so short a period of time and they set an agenda for years to come.

Part 1 of the report describes these events and assesses their effects.

Part 2 turns to the UK with a fresh evaluation of the manufacturing situation based upon the 4 key indicators used in the 1992 report [1], namely:

- *Output*
- *Productivity*
- *Share of World Trade*
- *Import Penetration and balance of payments.*

---

* Referred to subsequently as UK-SOS. A copy of the Executive Summary is included at Annex A.

** "Small and Medium-Sized Manufacturing Enterprises - A Recipe for Success", The Institution of Electrical Engineers, July 1993 (available).

*** The EC is still the only official organisation for legal purposes but post-Maastricht, EU is increasingly used and is in this report.

Plus an appraisal for the UK of the problems and opportunities of the GATT round.

Part 2 then goes on to describe the numerous changes in the UK domestic factors which influence future performance. Again, the same categories are used as in the previous report:

- *Education and Training*
- *Research and Development*
- *Innovation*
- *Management Style and Quality*
- *Size of companies*
- *Finance, investment and Government policy*

Part 3 of the report then summarises the challenges for the UK and the set of responses if these challenges are to be met successfully by our manufacturing industry.

The report concludes with just four suggested national initiatives, one each under the headings of Management, Technology, Finance for Industry, Innovation - all underpinned by Education and Training.

1.2   "Manufacturing"

There are agreed international classifications for all sectors of economic activity covering both "visible" and "invisible" trade. All the same, the word "manufacturing" tends to be used rather loosely in the literature.

In this report, the emphasis is firmly on manufacturing in the electrical, electronic, mechanical, chemical and aerospace industries. Broader trade figures are also quoted and care has been taken to ensure that all international comparisons are, as far as possible, on a like-for-like basis.

PART 1

INTERNATIONAL

2. THE EUROPEAN COMMUNITY (EC)/EUROPEAN UNION (EU)

2.1 General

The EU is one of the world's most prosperous and influential trading blocs; but it faces many economic problems including the recession, with its consequent unemployment; high costs; obstacles to restructuring; a lack of competitiveness compared with the USA and Japan; the challenge of Asia and the mismatch between eastern European and western European wages and living standards.

In total, European economies are gradually becoming more integrated and the European Commission has ambitious plans to develop transport, telecommunication and energy networks, spanning the continent. As yet, there is no agreed plan on how the cost of ECU 400bn will be financed. Also the EU faces massive problems in competing in technology and innovation with the USA and Japan.

The UK and the EU generally, has higher manufacturing costs per hour of labour than either Japan or the USA (Fig. 1). In fact, the EU also works the shortest hours and has the longest holidays.

Total labour costs in 1992
Wage and non wage costs (US=100)

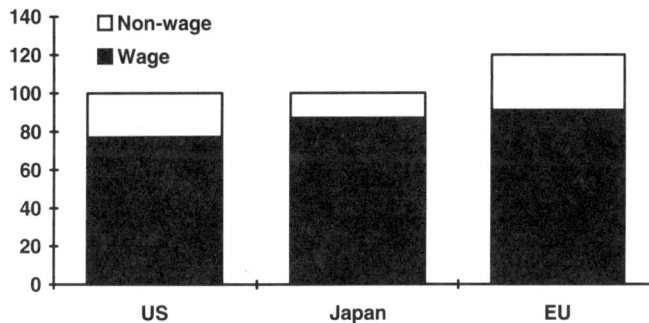

Source: US Bureau of Labour statistics

Long term unemployment 1991
Percentage of total unemployment

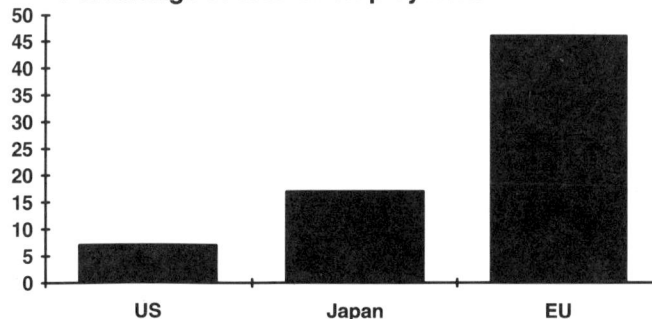

Source: OECD

Figure 1

3

Fig. 2 depicts the surplus or deficit each EU country has within the EU trading bloc. Fig. 3 shows that overall, the EU continues to show a huge trade deficit with the rest of the world. Moreover, economists are gloomy about the long-term trends for European competitiveness. According to OECD estimates, EU countries share of world markets for manufactured goods fell by 6.3% in 1991 with a further fall in 1992.

EU Members' trade surplus/deficit with rest of EU 1991

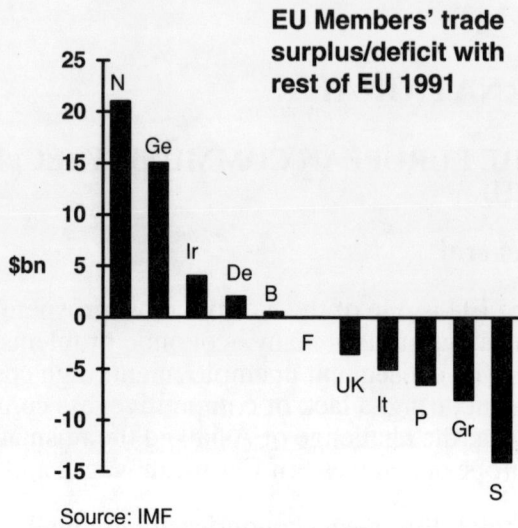

Source: IMF

Figure 2 EU's Trading Patterns

The automobile industry, to take a fairly representative example, is suffering from an explosion of capacity world-wide which, coupled with falling demand and relatively high costs, point to a crisis. In Europe, this is being met by proposed mergers and by restructuring of commercial plans, including the recent decision by VW to reduce investment in Skoda, the "quality" producer in eastern Europe and to close the SEAT plant in Barcelona. The EU total sales of 13.5 million cars in 1992 reduced to no more than 11.5 million in 1993 (Fig. 4).

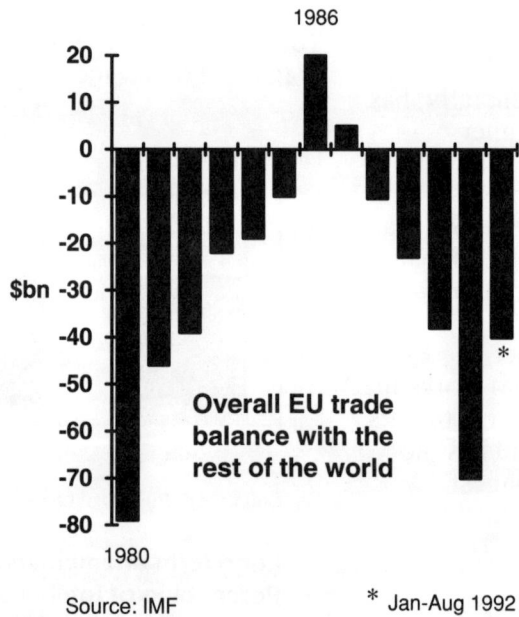

Overall EU trade balance with the rest of the world

Source: IMF   * Jan-Aug 1992

Figure 3 EU's Trading Patterns

In the long run, it is likely that there will be room for only three or four volume car producers in Europe and the recent buyout of Rover by BMW is part of a consolidation process leading in that direction.

In common with the rest of manufacturing industry and indeed service industries, higher productivity in automobile manufacturing is leading to a reduced work-force and consequent unemployment, a paradox which democratic market-led societies have yet to resolve.

2.2 Germany

Germany, generally reckoned to be the power house of the EU, faces special problems. What is happening in Russia has far more impact on Germany than it has on Britain. (German exports to Russia exceed the next two countries put together.) Also, the expense of reunification is producing social divisions and a bitterness which threatens unity.

Consequently, policy in Germany is currently dominated by an overriding concern to assist the rapid adjustment of the economy in the new eastern *Länder* to the needs of a market economy.

West European Car Sales

* Includes E Germany from Jul 90

** G M forecast

Million

■ Japanese share □ Total sales

Source: Automotive Industry Data

West European car market share 1992 total 13.5m

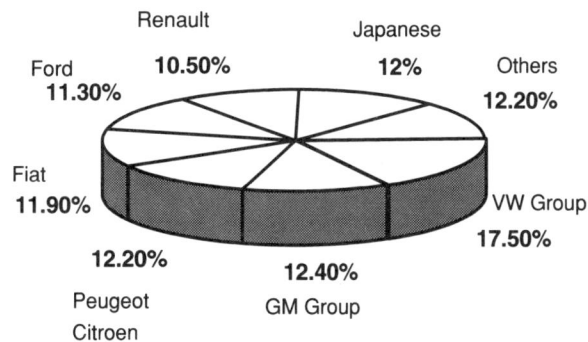

Renault 10.50%

Japanese 12%

Ford 11.30%

Others 12.20%

Fiat 11.90%

VW Group 17.50%

Peugeot Citroen 12.20%

GM Group 12.40%

Figure 4

According to a 1992 OECD report[3], one of the main thrusts is towards the establishment of diversified and efficient Small and Medium-Sized Enterprises (SMEs), together with investment in new and competitive jobs in both manufacturing industry and services.

By the end of 1993, the net results of policies to stimulate the East German economy had resulted in a situation where East German production was increasing, while that of West Germany was falling. These political and social problems for Germany are superimposed on a range of new economic problems.

To take just two just examples, the chemical and steel industries : Hoechst AG, Europe's largest chemical group not only announced disappointing results for 1993, but stated that capacity utilisation had dropped and that sales on the German market were down 15% on 1992. The other two of the German "Big 3" - Bayer and BASF have also reported a drop in profits. Bayer has warned that European politicians do not understand how cut-throat international competition is becoming.

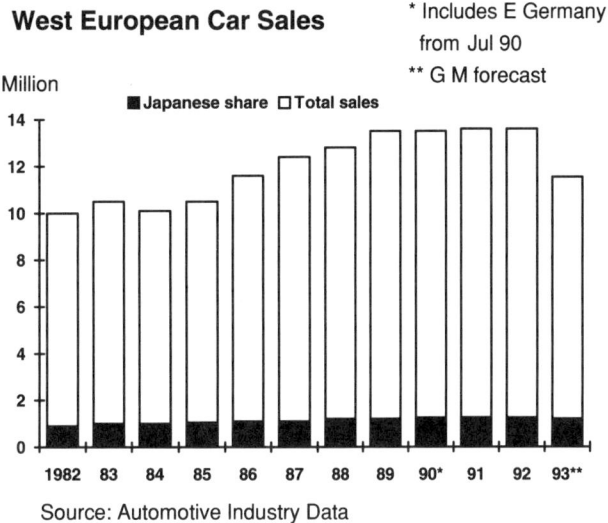

The chairman of Hoechst has stated that he sees this not just as a cyclical downturn, but as an indicator of the need for real structural change. What has happened to Hoechst - recession, high labour costs, extreme regulation, the need for restructuring - has been happening to most of German industry. Germany's costs now are probably the highest in the world, offsetting its traditional manufacturing efficiency.

Taking steel as a case history, over-capacity in eastern and western Europe, as well as the USA, has led to immense strains within the industry, particularly in Germany which is, by far, western Europe's biggest producer (Fig. 5). In 1992, world steel production was 604m tonnes, while consumption was some 50m tonnes less. The USA has placed penalties on steel imports and, to the fury of eastern Europe (where steel output has virtually halved), the EU has imposed anti-dumping measures. Imports from eastern Europe in 1992, although increasing, accounted for less than 3% of the EU market.

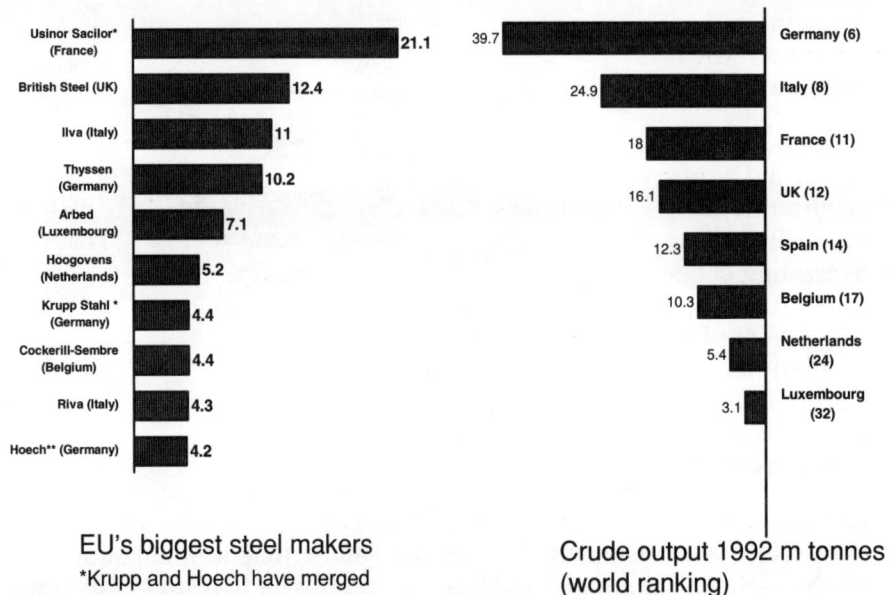

EU's biggest steel makers		Crude output 1992 m tonnes (world ranking)	
Usinor Sacilor* (France)	21.1	Germany (6)	39.7
British Steel (UK)	12.4	Italy (8)	24.9
Ilva (Italy)	11	France (11)	18
Thyssen (Germany)	10.2	UK (12)	16.1
Arbed (Luxembourg)	7.1	Spain (14)	12.3
Hoogovens (Netherlands)	5.2	Belgium (17)	10.3
Krupp Stahl * (Germany)	4.4	Netherlands (24)	5.4
Cockerill-Sembre (Belgium)	4.4	Luxembourg (32)	3.1
Riva (Italy)	4.3		
Hoech** (Germany)	4.2		

EU's biggest steel makers
*Krupp and Hoech have merged

Crude output 1992 m tonnes (world ranking)

Figure 5 European Steel Profile of Production

Source: International Iron and Steel Institute

Japan too has felt the severity of overproduction. Its five biggest steel makers have all announced losses for the year ending in March 1994. They are reducing capital investment and shedding jobs.

The EU itself took steps at the end of 1993 to reduce production by 30m tonnes annually. However, despite the protests of private producers, the state subsidised companies will reduce by only 5m tonnes and they will, in a politically motivated agreement by the industry ministers, continue to receive an annual subsidy of ECU 6.8bn. This leaves 25m tonnes to be shed by the unsubsidised producers who may have to close plant which is more efficient than in the state subsidised sector. There is considerable doubt whether reductions on this scale will be achieved. In a further controversial move the ministers approved the building of a hot strip mill by Ekostahl the struggling East German producer sending reverberations round the whole West German industry.

Amid this disarray, British Steel is one of the few producers to be trading profitably with pre-tax profits of £27m in the first half of 1993-94 and a creditable 56% home market share.

A further illustration of Germany's difficulties is that a new compromise pay deal for workers in the engineering industry was only achieved after employers threatened to reduce costs by the unprecedented cancellation of existing contracts for pay and holidays; and in return the union threatened a strike.

The scale of the problem is reflected by the difficulties of Metallgesellschaft, the metals, mining and industrial group with an annual turnover of £10bn and 250 subsidiaries. It is Germany's 14th largest company (founded in 1881 by an Englishman) and now employs 58,000 people. Late in 1993 it narrowly averted bankruptcy by means of an international banking deal which has been called the largest rescue operation since Dunkirk. The Metallgesellschaft case has prompted speculation that the German system of supervising boards with the role of shareholders reduced, compared to the UK and the USA shareholder system, is a weakness which could endanger the fabric of the German economy.

The German Government has now taken the lead in a national discussion based upon its Rexdrodt report published in September 1993 on the theme "safeguarding Germany's future as an industrial nation". It may seem oddly familiar to us in the UK that it deals with such topics as the growing burden of the welfare state, excessive public sector involvement in the economy, resistance to change, suspicion of high technology and the inflexibility of management. The report also dwells upon the high cost of labour and lack of competition in key parts of the German economy. It calls for a return to the "old values" of initiative, thrift and competitiveness. Among the challenges it perceives are the need to invest more state money into new technologies and to create an additional five million jobs by concentrating more on vocational and technical training, rather than on academic education.

Gunther Rexdrodt, the German economics minister, now forecasts a growth of 1.0-1.5% in 1994 and plans to deregulate the labour market, promote investment by small and medium-sized enterprises and to accelerate privatisation. All this seems to follow the UK example, but will provide even more competition.

2.3 Germany and France

The age-old antagonisms have been supplanted by the Franco-German economic axis in the EU with close collaboration at many levels. For example, French and German bureaucrats spend time in each others' civil service colleges and administrations helping to minimise differences between France's centralised system and the looser federal approach in Germany. Each has become the other's major trading partner (Fig. 6).

France Total Trade 1992 (%)   Germany Total Trade 1992 (%)

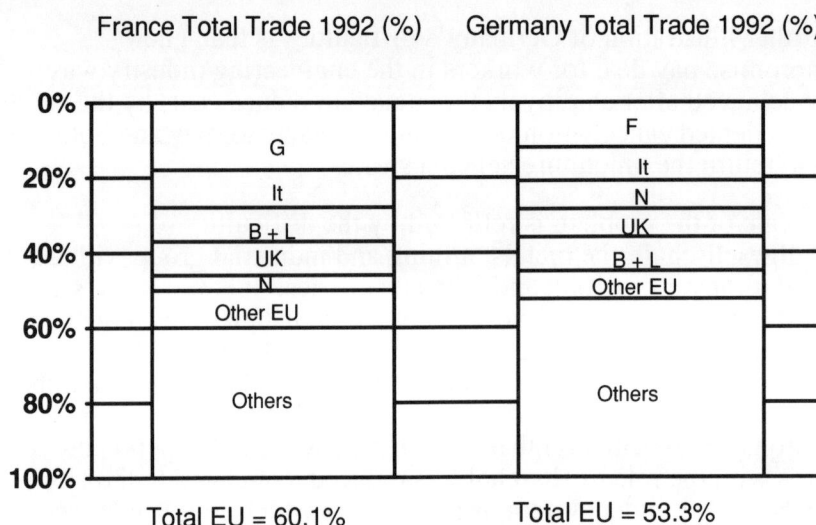

Total EU = 60.1%   Total EU = 53.3%

Source: IMF DOT

Figure 6 France and Germany: each others main partner

The recently announced strategic alliance between their state-owned telecommunication companies is indicative of the increasing industrial co-operation between Germany and France. France Télécom and Deutsche Telekom are the second and third largest international operators after AT&T of the USA. Subject to approval by the European Commission, the venture will be the first stage of a formal collaboration aimed to develop a "state of the art" European telecommunications network. The alliance may be interpreted as a defensive move, following those formed by AT&T with Asia-Pacific operators and by BT with MCI, the main rival of AT&T in the USA. Naturally, BT has concerns that the new Franco-German venture could be anti-competitive and a block to market forces within Europe. But the development does emphasise the way the two largest powers in the EU are now co-operating.

8

The relationship has changed somewhat since the re-unification of Germany and strains exist in such matters as the attitude towards GATT, policy on Yugoslavia and the enlargement of the EU. There is particular friction in the attitude towards monetary union. But the cohesion of the EU depends fundamentally upon the strong economic bonds between these two centrally placed countries in Europe. Although industrial production dropped in both by about 3% in 1993 and recovery will be slow, their combined manufacturing strength is a prime source of European development and will provide continuing challenges and opportunities for UK manufacturers.

2.4 EU Research

Scientific and technical research within the member states is perceived by the EU as a key factor in responding to the challenge posed by the USA and Japan. The standard of the technology on offer has become a crucial factor in international competition. In a world where production and innovation cycles are getting shorter, where R & D costs are rising and in which hi-tech sectors are increasingly interdependent, it is no longer possible to maintain the competitiveness and export capacities of Europe's economy within the traditional, nationally-based systems.

Accordingly, collaborative research between member states has gradually been increasing over the past decade. Although still small in terms of national budgets, it has risen steeply from 2% of the civil R & D budget of the 12 member states in 1980 to 5% now. The EU approach is based mainly on the phases of the "Framework Programme", concerned with pre-competitive research. The general philosophy is that of "subsidiarity" so that Framework resources are devoted to programmes that can be carried out more rationally, more cost effectively and more efficiently at European level and which promise real added-value as a result of cross-border collaboration.

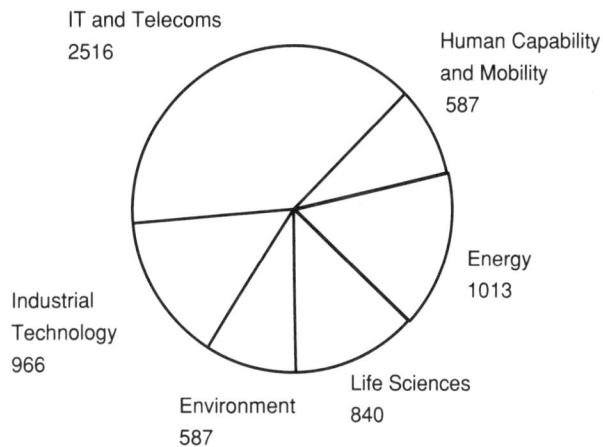

IT and Telecoms 2516

Human Capability and Mobility 587

Energy 1013

Industrial Technology 966

Environment 587

Life Sciences 840

Figure 7 EU Third Framework Programme (1990-94) Total 6600 ECU

The first Framework programme was in 1984-87, the second 1987-91 and the third 1990-94. Fig. 7 shows the expenditure profile of the third programme totalling ECU 6,600 million.

For the UK there were more than 6,200 links established with other countries between 1980 and 1991, the greatest numbers with establishments in France (1,408) and Germany (1,187).

The latest development is a proposal for the fourth Framework programme to run from 1994 - 98 and totalling ECU 13,100 million which includes the related work of the five Joint Research Centres (JRCs) operated by the EU at Ispra in Italy, Karlsrühe in Germany, Gael in Belgium, Petten in the Netherlands and Seville in Spain. Together the JRCs employ about 2,000 staff.

As automation in industry progresses and manufacturing employment consequently heads downward following the example of agricultural employment[4], so R & D is destined to become even more important if the EU wishes to keep pace with the USA and Japan. Some of the key areas are magnetic resonance imaging, neural networks, gene splicing, superconductivity and communications satellites.

2.5   The Enlargement of the European Union

On 1st January 1994 the world's biggest free trade market was formed, stretching from the Mediterranean to the Arctic Circle. The European Economic Area (EEA) extends the single market aspects of the EU to the European Free Trade Area (EFTA) countries, except Switzerland, that is, to include Austria, Finland, Iceland, Norway and Sweden.

The removal of many trade and finance barriers will provide opportunities for UK manufacturers to build upon the £5.7bn exported to the EFTA countries last year.

Austria, Finland, Norway and Sweden are in an advanced state of negotiations to become full members of the EU and this could happen as soon as 1995. This will further increase the potential strength of the EU as a global trading bloc, but will add complexity to the harmonisation process.

The North American Free Trade Area (NAFTA) between the USA, Canada and Mexico also came into force on 1st January 1994 (section 4.2). Comparing the two shows the EEA to advantage.

	EEA	NAFTA
Population	372 million	364 million
GDP	$7,500bn	$6,200bn
GDP/head	$20,160	$17,020

3.   EASTERN EUROPE

3.1   The unsettled scenario

Are the states of the former USSR, especially Russia, lurching towards stability or chaos?
What are the implications for UK industry and manufacturing?
Will the CIS group become major markets for UK goods?
Will they eventually export more to the UK?
Will they become serious competitors in our other export markets?

If the chaos scenario prevails, with secession of small groups from the major countries, accompanied by civil wars, it is difficult to see how the CIS will prove to be either a significant market or a threat to the UK economy.

Fig. 8 shows the drop in Russian industrial output and the pattern of oil and gas production. Such a steep decline does not apply to the "westernised" states such as Hungary, and Poland where, in spite of great difficulties, stable market economies are on the way to becoming established (Fig. 9). Since 1989, about $9bn of capital investment has gone into the countries shown, with the majority ($5.5bn) going to Hungary. Poland is planning a comprehensive motorway and fast rail system as one of the NS-EW cross-roads in Europe.

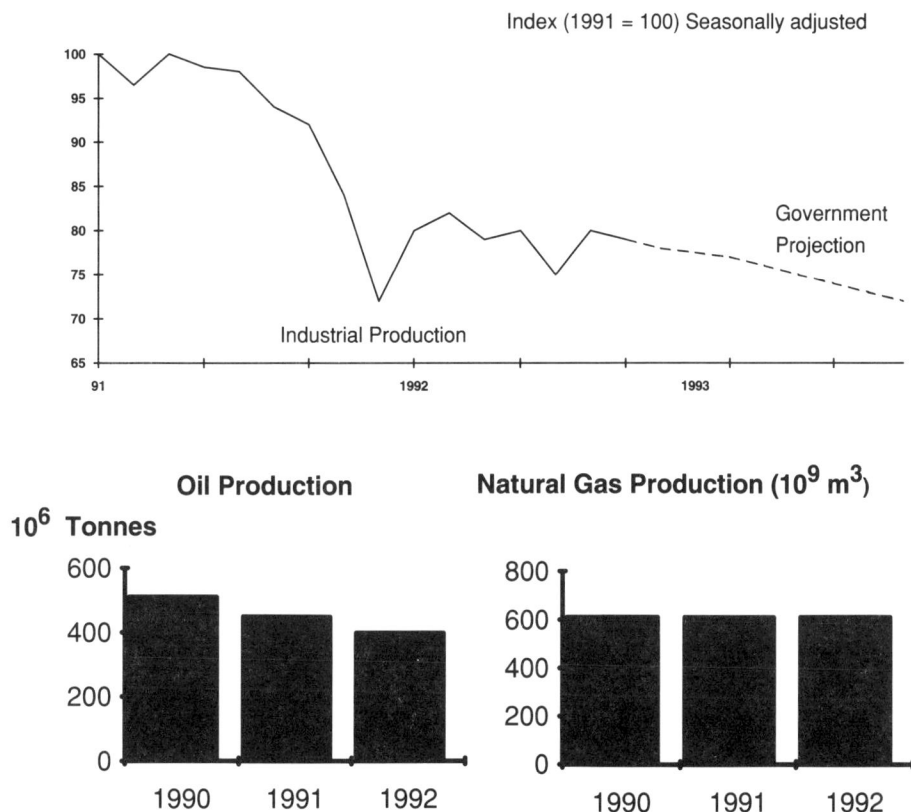

Figure 8 Russia's Decline

Source: Russian Economic Trends

11

GDP % real growth

Industrial Sales (1990=100)

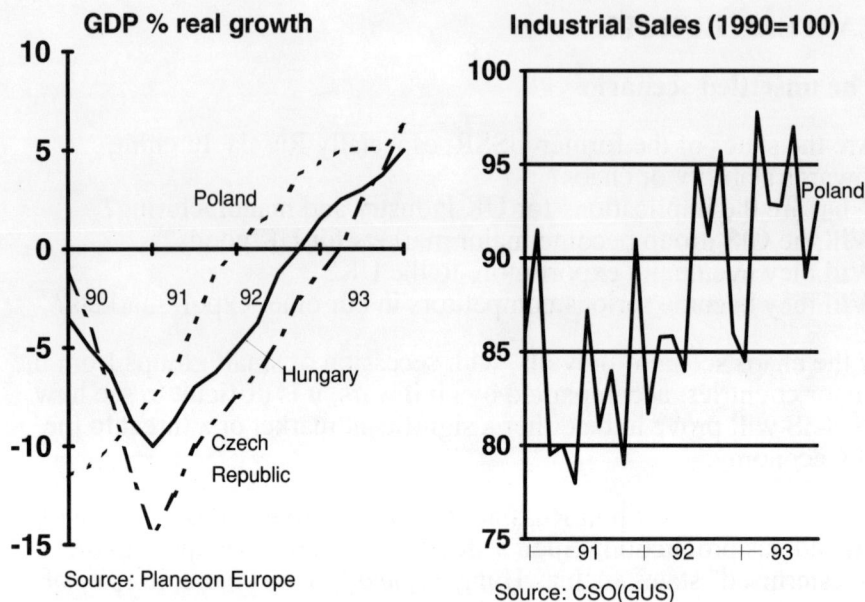

Figure 9 Poland's Path to Recovery

The same two countries plus Romania, Bulgaria and the former Czechoslovakia have reached agreements with the EU which commit the signatories to industrial free trade within ten years. Steel, chemicals, clothing, textiles and processed foods are excluded. The EU will continue to exert a degree of protectionism in these sectors to the annoyance of the eastern European countries. As the Common Agricultural Policy (CAP) already excludes most agricultural produce from eastern Europe, these restrictions are likely to prove a barrier to free-trade and better relations for many years to come.

Even in Russia itself the signs are not all negative. In spite of the struggle for power between the old and new guards, inflation has been heading downwards from 50% per month (a value which verges on hyper-inflation). It was seemingly under control in the last quarter of 1993, although political turbulence has now caused a further devaluation and inflation could be in the range 10% - 20% a month.

Privatisation has been making steady progress, so that by the end of 1993 more than one-third of heavy industry and the majority of small firms were privately owned. The basis of Russia's privatisation programme has been the voucher distributed free to all Russians at the end of 1992. An estimated 35m people are now share-holders who, hopefully, will be in favour of more privatisation and more democracy. There are, of course, still many anomalies - for example domestic energy prices are only 5% of world levels.

3.2 An example in telecommunications

Telecommunications provide a vital underpinning for a modern industrial culture so it is instructive to look at an aspect of that sector in eastern Europe. Table 1 shows the telephone line situation in comparison with the western European average of 430 lines per 1000 people.

Table 1 Telephone Lines - Eastern Europe

	lines/1,000 people	Additional lines (m) Required to reach 350 per 1,000 people	Population m
Baltic States	225	0.99	8
Bulgaria	256	1.05	9
Russia	153	29.31	149
Ukraine	154	10.19	52
Czech Republic/Slovakia	243	1.10	10
Hungary	113	2.45	10
Poland	94	9.78	38
Romania	103	5.73	23

A total of more than 60 million additional lines are needed which, at an average of $2,000 per main line, requires an investment of some $120bn or about $15bn per year to reach the target by the year 2000. Also, this calculation neglects the fact that much of the existing equipment is outdated and needs replacement. Western firms can find themselves in a dilemma, as they do not want to be left out of potentially profitable development, but at the same time do not wish to take large risks. However, there have already been some commercial agreements made in Hungary and Poland for fixed links and mobile phones. At the end of 1993, Deutsche Telecom and Ameritech of the USA clinched Eastern Europe's biggest single privatisation deal, agreeing to pay £600m for a 30% stake in Matav, Hungary's state telecommunications company. Tenders have now been accepted from foreign-led consortia to operate local telephone services in 15 of Hungary's 54 telephone districts.

The telecommunication gap is representative of many other sectors of the eastern European economy. Run-down factories, dangerous mines, oil-fields in disrepair, inadequate distribution systems and all kinds of environmental and ecological disasters are characteristic of the problem. To overcome these difficulties, the investment in money and know-how will be huge and should provide many opportunities for UK manufacturers as the political scene settles down.

3.3 The Way Forward

Russia has a highly educated population and, if the trappings of communism are finally discarded, could easily be in a position to challenge the west in the manufacturing arena. After all, a country which has the capacity to design and manufacture advanced aircraft and space vehicles, cannot be dismissed lightly. The talent is there to be harnessed. In spite of the obvious risks, UK industrial firms should be carefully cultivating links behind the former Iron Curtain as friendly partnerships with the new private industries could become important later on. The opportunities could be huge because there is a great deal of leeway to be made up and investment is relatively cheap by western standards.

The way forward for the UK in the vast new markets of the CIS will often be via collaborative agreements on manufacturing using British expertise, much as the Japanese have done here. For example, an agreement has been signed between UK CADCAM supplier Delcam and the Saratov Aviation Plant. Delcam will supply workstations to the 13,000 employee factory located about 800km south west of Moscow. Delcam has six sales/training offices in Russia and the Ukraine. Also, a joint training office is to be formed in Moscow with the National Institute of Aviation Technology.

Interested parties in this software now include Russian refrigerator, motorcycle and valve manufacturers.

But there is much competition : top Russian and USA companies are to develop a new range of drilling rigs for the Russian petroleum industry and the joint venture will operate in what is potentially the biggest market for oil equipment in the world.

4. THE UNITED STATES

4.1 General Survey

Traditionally, in times of recession, the subsequent recovery has been led by the USA, the largest economy in the world, even adding all the present EU states together. As Fig 10 shows, the USA has indeed shown signs of economic expansion while Germany and Japan slowed down in 1992 compared to 1991.

14

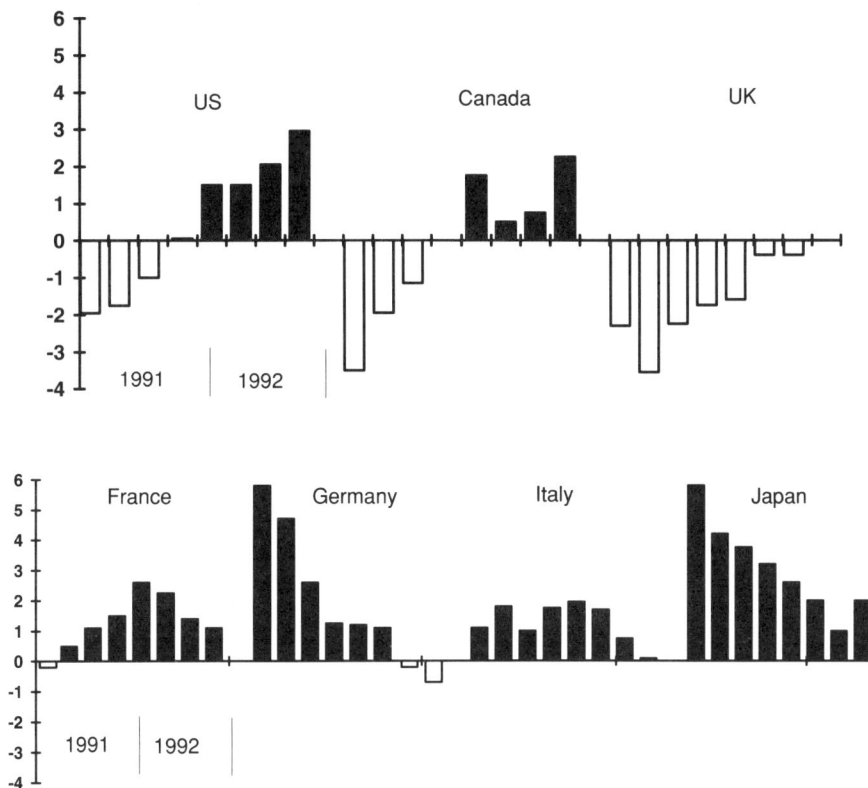

Figure 10 Real GDP Growth - Annual % Change

Source: Datastream

This time, however, there is more uncertainty about the sustained ability of the USA to exert its historic effect. In some respects the American dream may be failing. Although many more jobs have been created, they are not in the traditional middle-class areas of work and average real earnings have been almost stagnant. The national debt of some four thousand billion dollars (and growing) gives deep cause for concern. Huge losses incurred by IBM have also shaken confidence in the dominance of American business and technology.

Nevertheless, Compaq, Hewlett-Packard and Microsoft are still making money and the merger of Bell-Atlantic and Tele-Communications Inc., the largest cable TV company in the USA, will form a conglomerate with sales of $16bn a year. It will exploit the information revolution which could transform world communications and entertainment. This merger is so large that it could well have decisive leverage in setting future technical standards and patterns of inter-corporation joint ventures.

15

The American economy can be seen as amazingly resilient when it is remembered that the "peace-dividend" is resulting in huge defence cuts (Fig. 11) from 5% of GNP in 1991 to a projected 3% in 1998, although this is part of a long term trend from 12% in the 1950's.

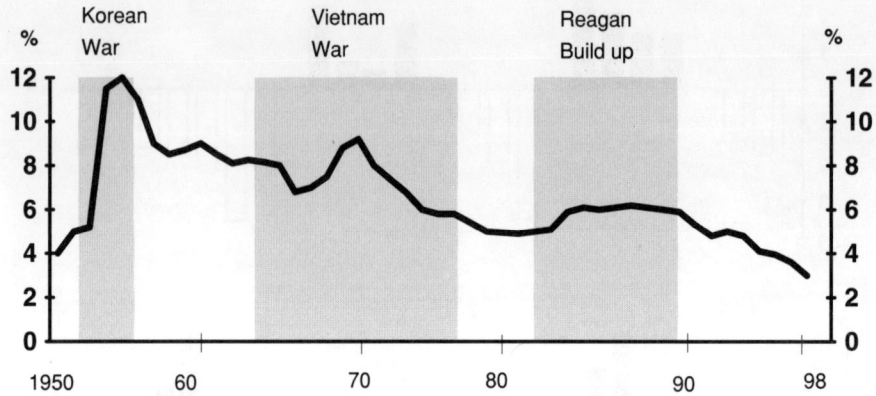

Figure 11 US Defence Outlays as % of GDP

Source: US Dept of Defense

Also, Europe is making inroads into the USA lead in civil aircraft manufacturing. Combined with the economic downturn, this has meant that Boeing has had to cut production of its entire family of airliners, so that deliveries are down from 446 in 1992 to 350 in 1993 and perhaps 258 in 1994. Nevertheless, Boeing is fighting back with the new 777. Its efforts are paradoxically being aided by British Airways which has ordered 15 aircraft with GE engines, so angering Airbus and Rolls Royce simultaneously. Rolls Royce had hoped to fit its Trent engine which has been selected by the Brazilians for three 777s. Also, a jointly developed engine by Rolls Royce and BMW will power the new McDonnell Douglas MD-95 from 1998.

On the broader industrial front, the USA trade deficit with Japan remains huge but, in spite of this, there is heartening news, not just for the USA but for the rest of the world too. The American GDP figures for 1992 lead the way with an improving annual rate as shown in Fig 10. This trend continued into 1993 and it is believed by many forecasters that a growth rate of 3% is now assured some way into the future. Others are nervous that Mr Clinton's tax increase - the largest in American history - will reduce the GNP in 1994, 1995 and 1996. The business community is concerned that the financial burden of the proposed new health insurance will adversely affect manufacturing costs and that the threat of price controls on pharmaceuticals will have an effect on other industries too.

These negative factors are balanced by plans for investment over the next four years, part of which are summarised in Table 2.

Table 2 USA Investment Plans 1993-97

	$bn
Infrastructure	48
Lifelong learning	38
Investment incentives	24
Transport	8
Environment	8
	$120bn

Additional funds for short-haul aircraft research, dual use technology for defence reinvestment, high performance computing and R & D tax credits are specific for 1994-97. The impressive goals of the total American programme are given in Annex B and should be taken seriously by the UK, because the Americans are in the habit of meeting their national targets and finding the resources necessary to do so. For example, in the key field of information networks, Dr Frank Carrubba, chief technical officer of Philips, says that the Americans are 2 years ahead. Annex B gives a vision that the UK would do well also to adopt.

A key passage runs : " American technology must move in a new direction to build economic strength and spur economic growth.................recognising that Government can play a key role in helping private firms develop and profit from innovations".

More recently, President Clinton has unveiled plans to double American exports to about $1,000bn annually by the end of the decade, partly by relaxing export controls on high technology items hitherto covered by licensing rules in the cold war situation. This move will inevitably produce more competition for the UK.

4.2   The North American Free Trade Agreement (NAFTA)

This important agreement between the USA, Canada and Mexico will, when finally put into operation, produce a common market larger than the present EU in terms of population and GDP (Fig. 12). Over the next 15 years it is planned to abolish all trade barriers across almost the whole of North America from Alaska to Yucatan. However, NAFTA is a marriage of disparate partners, much more so than the EU.

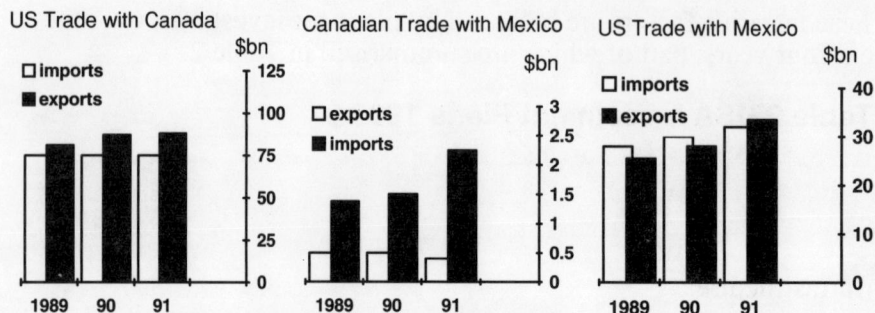

	CANADA	US	MEXICO
Population m	26.6	151.5	86.2
GDP $bn	570	5,392	238
GDP/capita $	21,418	21,449	2,490

Figure 12 The North American Market US $bn

Source: Commerce Department Statistics Canada. Organisation for Economic Co-operation and Development World Development Report 1992

While free trade already substantially exists between the USA and Canada which have similar cultures, Mexico is markedly different in its political rule and GDP per capita.

Doubts about the effect of NAFTA on the USA are reflected in the narrow margin in favour of acceptance in the Congress, but the basic message behind NAFTA is the support of the USA for freedom of trade, so putting it in a stronger position in its developing relationships in Asia and in exporting to the EU.

4.3 Latin America

In addition to NAFTA, there is a lesser known free trade agreement in the Americas in which the USA, although not a member, has a geographic interest. This is MERCOSUL, signed in 1991, which aims for the free movement of capital, goods and services between Argentina, Brazil, Paraguay and Uruguay. Because of the disparate, though sometimes complementary, nature of the four economies, rapid progress is unlikely until the turn of the century.

4.4 The USA/Japan Trade Dispute

The recent decision of President Clinton to lift the 20 year trade embargo on Vietnam, now seen as a possible prime market, is just one of many USA policy changes to bolster trade links in the Pacific Rim (see Section 6). This move may herald a de-coupling of trade agreements with human rights questions and so may add competitive pressures.

By far the largest developed economy in South East Asia is Japan, whose closed domestic market is shown dramatically in Fig. 13. For their part, the Japanese maintain that comparisons with other countries are misleading because Japan is more competitive.

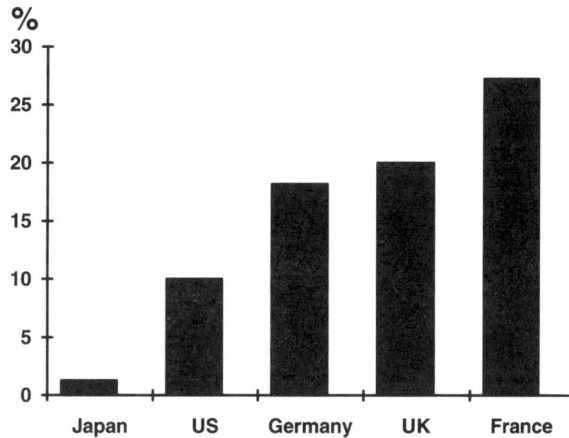

Figure 13 Foreign Companies' share of domestic sales 1986 (%)

The problem is not new. Trade disagreements of various kinds have marred USA/Japan relations for most of this century. The situation is a continuing problem for the USA, which can hardly be accused of lacking in competitiveness, yet has an annual $50bn trade deficit with Japan. The debate outside Japan centres on the best way of achieving change, either by managed trade sector-by-sector or by "strong-arm" measures to achieve a general relaxation of barriers.

The classic instance of the "managed-trade" approach is the 1986 agreement on microchips, that by the end of 1992 American firms would be expected to gain 20% of the Japanese market. But such agreements can serve to limit penetration as well as encourage it. In competition with the Japanese, the USA has 50%, not 20% of the European and Asian markets. In fact, by the end of 1992, American firms had only achieved 16% of the Japanese market (Fig. 14) and despite a more recent rise to 18%, this is regarded by the USA negotiators as an agreement not kept.

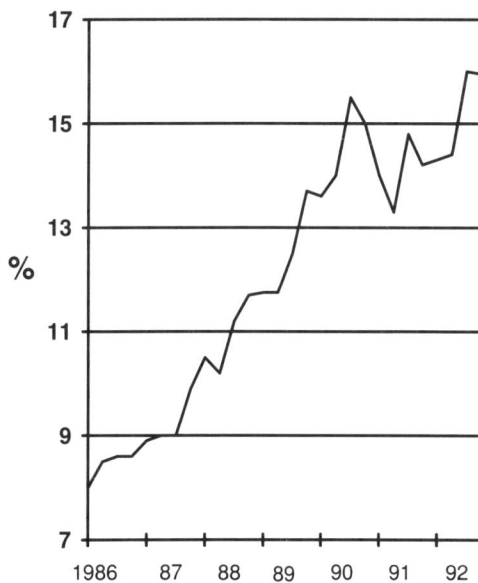

Figure 14 Foreign Market Share in Japan as defined by the 1991 US-Japan Semiconductor Agreement

Source: SIA and USTR

However, judging from the breakdown of recent negotiations, it is unlikely that Japan will ever again agree to a numerical goal of any kind. So policy in the USA now is to exert

19

maximum pressure to achieve a general relaxation of Japanese barriers in key trade areas such as telecommunications, cars, medical equipment and insurance. Otherwise, trade sanctions are threatened on Japanese goods.

The USA had hoped for EU co-operation in its policies but, although the EU would like to see Japanese markets opened up, it has reservations on USA tactics.

Also, the USA is trying to force up the value of the yen to reduce Japanese exports but this is a double-edged weapon because it also lowers the cost of the huge volume of Japanese oil imports.

This is a game that the USA probably cannot win. The best that can be hoped for is an honourable draw, such as that represented by the March 1994 agreement between Motorola and IDO, the Japanese mobile phone operator, which gives the American company access to the Tokyo-Nagoya region but which, officially, at any rate, does not involve the Japanese Government.

The linked internal problems of the Japanese economy are covered in the next chapter.

4.5   The USA - Conclusion

While the latest figures show that in 1993 the USA economy grew by 2.9% compared with 2.6% in 1992, the uncertainties of NAFTA, the huge USA deficit, higher taxes and the cost to industry of the new health plan mean that there is still doubt about whether any recovery in the USA will be sustained, and if it is, whether it will lift the rest of the world with it. The success of the GATT round, on which the USA pressed hard, will be particularly significant in this respect (see Section 7.6) and the reductions in tariff barriers for agricultural and manufactured goods should assist the USA in its export efforts.

5.   JAPAN

5.1   General

Japan -the guiding light and focus of attention for all manufacturers. But is the Japanese economic miracle over? Since the war, the Japanese economy has grown at a rate unsurpassed in the history of the world, but the pace of growth has slackened and a recent OECD report [5] concluded that Japan is "undergoing a period of marked cyclical and structural readjustment to the excesses of the late 1980s boom". For example, the automobile industry, a good economic indicator, is slowing down with reductions in both car and truck sales. In February 1993, Nissan announced plans to shut a domestic production plant to save $2bn with the loss of 5,000 jobs. Profits dived at NEC, Toshiba, Hitachi, Mazda and Matsushita, while the telephone communications group NTT plans to cut 30,000 jobs over the next 3 years. Deep cuts in investment are widespread and not superficial. Evidence is accumulating of a great deal of hidden unemployment and also that the traditional Japanese attitude to a total work ethic is changing.

Japan has successfully weathered previous recessions but this time a combination of circumstances may prove to be more difficult to surmount. Production and GNP are almost stagnant (Fig. 15). Land prices have plummeted as has the stock market, with the Nikkei index down from 40,000 in 1989 to less than 20,000. These changes have caused a massive contraction in corporate and personal wealth which, in turn, has put enormous pressure on the banks with a sharp deterioration in their balance sheets. The financial system as a whole may be carrying a huge sum of bad debts which has not yet come to light.

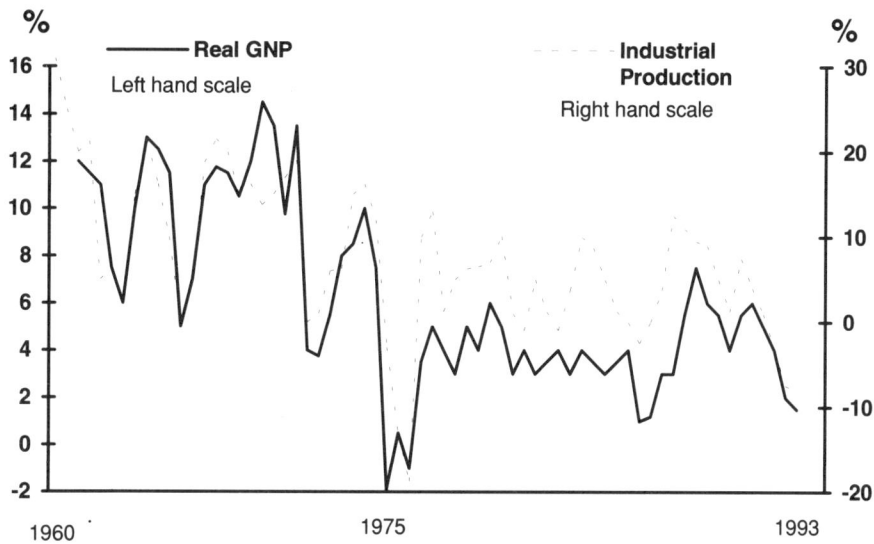

Figure 15 Japan's Decline % change year on year

Source: MITI

This is not just an external perception. According to the Japanese Ministry of International Trade and Industry (MITI) "The Japanese economy is in an adjustment phase with economic indicators pointing to a major slowdown". This is reflected for example in a reduction in job-openings - in 1993 manufacturing job offers fell 26% year on year, overtime hours reduced by 10.9% and the operating profit to sales ratio has fallen from 5.47% in 1984 to just 2.88% last year.

On the positive side, Japan has tremendous industrial muscle and still enjoys huge favourable trade balances with the USA and EU. Its industries are taking vigorous steps to develop their global strategies, reduce manufacturing costs, trim capital spending and give greater focus to research and development. What is now being tested is Japan's total socio-economic capacity to change in the new economic climate. One of the unique features of the Japanese scene is the existence of cross-share holdings which can allow long planning horizons, but may inhibit competition and restructuring.

21

The Japanese Government has put forward an economic package consisting of some short-term pump priming and long-term reforms to steer away from recession. To date these have failed to inspire a great deal of confidence. All the signs point to a long, hard recovery.

Japan now recognises the need to share internationally the costs of developing new technologies that are not considered core activities and of the need to move manufacturing offshore to be near markets or to take advantage of lower cost labour while retaining the added value content.

5.2 Japan and Asia

For Asia as a whole, Japan plans to move centre stage and is in a strong position to do so. Its bilateral ties are as complex as the individual nations are diverse (Table 3) but a clearer "pan Asia" policy is in the offing now that memories of wartime occupations are fading. South Korea, Indonesia, Malaysia, Vietnam, Brunei, Thailand and Burma are all targets of Japanese diplomacy. Asian markets already account for 41% of Japan's total trade compared with 30% going to the USA. Five years ago both regions had 35% and Asia is regarded as a bright spot amid the gloom of international recession. "As the Asia-Pacific region gains further influence and moves towards becoming an open, pluralistic region with a central role in the global community, the Japanese people must assume a heavy responsibility" says a prime-ministerial advisory panel on Japan and the Asia-Pacific region published in December 1992.

5.3 Japan and China

As Japan's biggest neighbour, China has special significance. Japan is trying hard to improve the relationship and in 1993 its investment in China doubled. In fact after the USA, China was Japan's second largest trading partner in 1993. In the first half of the year, Japans' exports to China rose 51% compared with a year earlier. Exports of cars and television sets increased sixfold and fourfold respectively. Exports of motorcycles, communication systems, steel and construction equipment all doubled.

The Chinese market is clearly discerned as providing excellent opportunities for export growth, yet there is an ambivalence in the relationship. Although the Chinese market provides vast possibilities for Japanese industry (and for the UK by the way) improvement of Chinese goods, often assisted by Japanese technology, is perceived as a long term threat to Japan.

Flowing the other way, Chinese exports to Japan increased by 15% in the first half of 1993, including a tripling of office machinery made in Japanese-run joint venture factories. It is not difficult to appreciate Japanese fears of growing Chinese political and military influence in the region and hence industrial dominance.

6. THE PACIFIC RIM

6.1 A Survey

The Pacific Rim countries are situated roughly on a huge circle about 8,000 miles in diameter with its centre in the middle of the Pacific Ocean, slightly offset from the equator. Their populations and related GDP per capita are shown in Table 3 (overleaf).

A comparison of some Pacific Rim countries with other "third world" economies is shown in Fig. 16 in terms of growth rates. Korea, China and Thailand in particular have enjoyed huge growth rates since 1965 - though they had much catching up to do compared with the developed nations. The same three countries show 10% or more annual growth of export volume since 1980 (Fig. 17).

Malaysia, not shown in Fig. 16, is another expanding economy with 8.2% growth in 1993 - a rate maintained for the previous six years. It aims to become fully industrialised by the year 2020. The rate of expansion is such that the capacity of the infrastructure - roads, ports, power is under considerable strain but this is a problem worth having! The Pergau dam, which has recently hit the headlines in the UK, is part of the infrastructure programme.

The major economic players are the USA, Japan and China. The many small and medium-sized countries have a mix of traditions and occupy the whole of the political spectrum. They vary from vigorous manufacturing economies to largely agricultural subsistence societies.

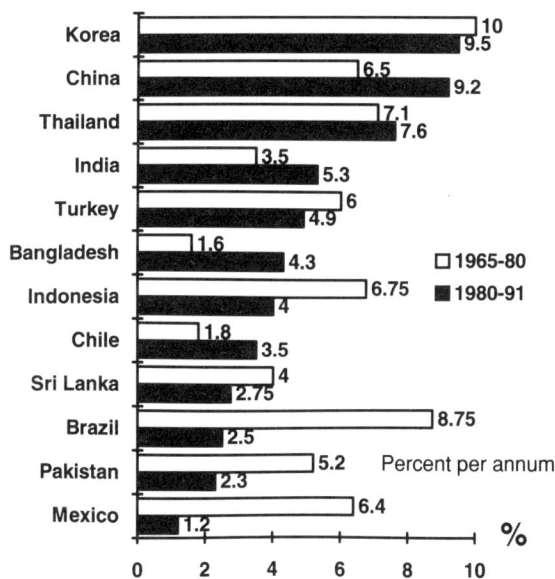

Figure 16 Growth Rates of GDP

On the macro-economic plane it is becoming clear that the eyes of the USA are, to some extent, turning away from Europe and focussing more on the Pacific Rim countries. The USA may well have ambitions to lead and co-ordinate the economic development of the region so the countries of the EU will need to be vigilant if they are to protect, let alone expand, their exports and exert influence among this important group. This again may bring the USA into further trade conflict with Japan which has similar ambitions on its side of the Pacific Rim.

23

Table 3 The Pacific Rim Countries (Population m/GDP per capita, USA $)

Country	GDP per capita	Population m	Country	GDP per capita	Population m
Russia	?	148	Canada	21,000	27
Japan	23,800	124	USA	21,000	252
South Korea	6,500	43	Mexico	2,500	86
China	400	1,143	El Salvador	1,000	5
Taiwan	8,800	21	Nicaragua	900	4
Hong Kong	13,400	6	Costa Rica	1,800	3
Vietnam	200	69	Panama	1,800	2
Philippines	800	61	Colombia	1,300	34
Singapore	12,400	3	Ecuador	1,100	10
Thailand	1,200	58	Peru	940	22
Malaysia	2,500	18	Chile	3,000	13
Indonesia	500	183			
Australia	14,000	17			
New Zealand	12,000	4			

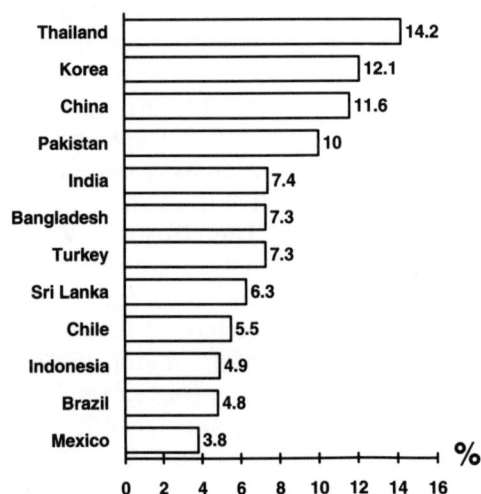

Country	Growth %
Thailand	14.2
Korea	12.1
China	11.6
Pakistan	10
India	7.4
Bangladesh	7.3
Turkey	7.3
Sri Lanka	6.3
Chile	5.5
Indonesia	4.9
Brazil	4.8
Mexico	3.8

**Figure 17 Growth of Export Volume
1980-91 (Average annual rate %)**

Education indicators for many of these developing countries are shown in Table 4.

Table 4 Education Indicators for some Pacific Rim and other Developing Countries

	Literacy % of Total	Secondary School Enrolment %		Tertiary School Enrolment %	
	Population	1965	1989	1965	1989
South Korea	95+	35	86	6	38
Taiwan	95+	33	95	7	33*
Thailand	93	14	28	2	16
Hong Kong	95+	29	73	5	13*
Malaysia	78	28	59	2	7
Singapore	95+	45	69	10	12*
Argentina	95	28	74	14	41
Brazil	81	16	39	2	11
Chile	93	34	75	6	19
Mexico	87	17	53	4	15
China	73	24	44	0	2
India	48	27	43	5	na
Indonesia	77	12	47	1	na
Philippines	90	41	73	19	28*

*1985

6.2   South Korea

The example of South Korea is instructive in the way it has developed its manufacturing industry. After numbers of social experiments

Korea has discovered that traditional economic fundamentals plus a huge investment in education and training (see its premier position in Table 4) coupled with industrial policies designed to promote particular sectors have been the keys to success.

Korea now has the world's largest car plant, shipyard, TV factory and steel mill, all privately owned and making money.

But, even the dynamic Korean economy is now showing signs of strain. One of the great advantages of these new economies is the low wages compared with those paid in the west. This has promoted exports and made it extremely difficult for developed countries to export to Korea on a competitive basis except for specialised equipment. However, there appears to be a medium-term inherent balancing mechanism at work because as economies prosper, workers apply pressure for an enhanced standard of living. In Korea this is already happening, as is indicated by the 130% increase in wages over the past five years.

Problems are also being caused by the very heavy concentration of power in the monster conglomerates such as Samsung, Hyundai and Daiwoo. The Korean Government has made several attempts to reduce the size of these "chaebol" which, on average operate in nine different industries. In the Government's view this harms their efficiency and international competitiveness. Financial inducements such as lifting borrowing restrictions and providing support for research and development in core business areas, are now being offered to induce the 30 largest chaebol to reduce their activities to two or three sectors only. Non-core industrial sectors will be subject to strict credit restrictions.

The Koreans are actively planning to become more independent of Japan and assisting the development of technology-orientated small and medium-sized enterprises. They are also devoting attention to improving infrastructure with highways of both motorway and information types. Some of those projects do indeed provide opportunities for UK and European companies. For example, GEC-Alsthom is in the running to supply equipment for the planned Korean high speed train, and has recently been chosen for a similar project in Thailand.

6.3   China

There is a saying that "China is a civilization pretending to be a state". Be that as it may, China is evidently the real rising power in South East Asia. It is almost impossible to overstate the future importance of the Chinese market and the importance of developing the involvement of UK enterprises.

Chinese industrial production is growing very fast (Fig. 18) averaging 15% annually since 1985 with individually owned enterprises growing fastest of all. Over 50 of China's largest enterprises are to be re-structured along the lines of Japan's successful "sogososha" trading conglomerates. The pace of privatisation can be judged from the fact that 15 years ago 90% of industrial output was from the state sector while the proportion is now 45% on a much larger volume of production.

Annual Growth by
ownership type %

Industrial Production
% Annual Growth

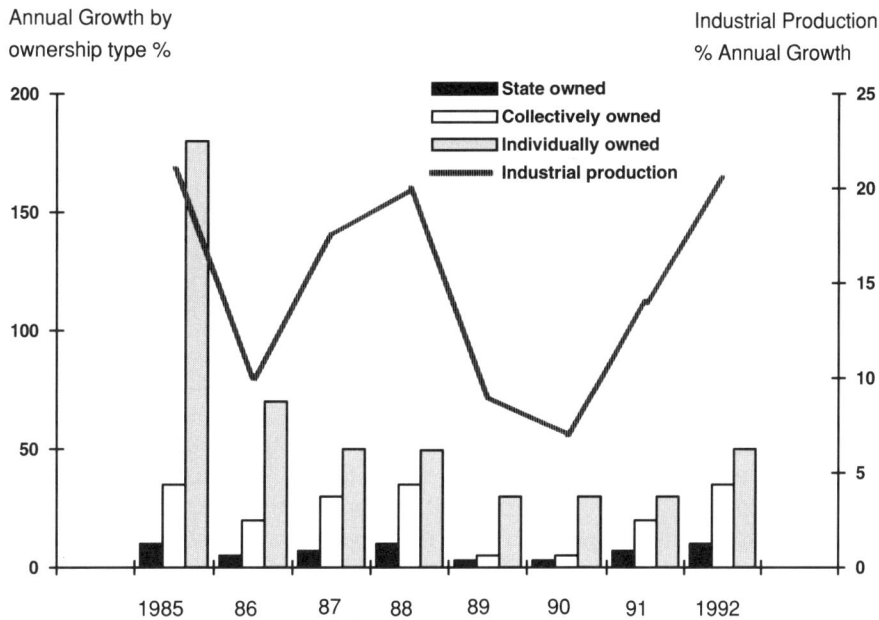

Figure 18 China's Economic Boom

China's positive trade balance with the USA, at $18bn in 1992 and probably $24bn in 1993 is second only to Japans   The economy is, in the phrase of the World Bank, "half-reformed". USA exports to China are mostly in sophisticated technology, capital equipment and aerospace and are steadily growing.

Whether growth at the same rate can be sustained is an open question but, with cheap labour and a growing technological base, China is well placed to expand its manufacturing if it can contain inflation, exert proper financial controls and smooth out the huge differences in prosperity between its provinces. New tax structures are being introduced, including an overhaul of income tax and the introduction of VAT. In these endeavours, China has an economic advantage over Russia in putting industrial regeneration ahead of political reform although, in the opinion of the author, the latter will inevitably follow the former to the benefit, for example, of Hong Kong.

Table 4, however, shows that China still has a vast gap to bridge in its educational system if it is to develop its economy on a par with the West. The GDP per head is only $400 per year but the potential is enormous. The Japanese, for example, are eyeing the prospective car market in China with anticipation.

There is now a rush of foreign investment into China with investment in 1993 exceeding $15bn. The Beijing Government approved some 63,000 projects, worth $83bn, with foreign participation nearly double the 1992 figure.

Many European companies are involved and the UK must not be left behind. Typical is Bayer which has signed an agreement with the Chinese Chemical Industry Ministry and is investing $200m as part of a drive to establish a comprehensive presence in China.

Foreign manufacturers are waging an intense battle to share China's booming telecommunications market. AT&T has a considerable presence and is negotiating to set up a research and development operation in the country. Siemens has a chip manufacturing plant near Shanghai and the Japanese multinational NEC produces semiconductors in Beijing. Northern Telecom is negotiating to make switching equipment in Guangdong province and also won a $160m contract to sell such equipment to four Chinese provinces.

China is now poised to take advantage of the latest technology. Not surprisingly, there are plans to develop an extensive optical fibre communications network.

The question that must be asked is whether the UK is paying enough attention to this major emerging nation, bearing in mind that the Chinese often impose a financial penalty from those late in the market, if others have taken the initial risks. Sino-British trade shows a healthy increase both ways but results in a deficit against us which has been growing steadily. In 1986 there was a £200m balance in our favour, but in 1993, a £600m gap the other way. This is simply not good enough.

China does face problems in its relationships with its neighbours. Although those with Japan are generally improving, it has territorial disputes with other neighbouring countries such as Vietnam, where there are arguments over oil - drilling rights in the Gulf of Tonkin and over the ownership of the Spratly islands in the South China Sea nearly 1,000 miles off the Chinese coast towards Malaysia.

Altogether there is a nervousness among all its neighbours about Chinese development and possible long term ambitions. This is made worse by reductions in the USA military presence which has been seen as a stabilising factor since World War 2.

A further complication is the possibility of the USA withdrawing "most favoured nation" status from China because of a lack of progress in democratic reforms and human rights. Should this happen, USA tariffs on Chinese goods would increase and Hong Kong trade would reduce by at least 10%, being the transit for a high proportion of Chinese goods. In fact, the Chinese warn that the cancellation of preferential trade access to the USA market would have devastating consequences for the Hong Kong and Taiwan economies and would limit access for USA business to China.

6.4 Asia-Pacific Economic Co-operation (APEC)

APEC was formed five years ago on the initiative of Australia. Unlike previous trade organisations in the area, it brings together governments. The 17 states now include Japan and China and also the USA, which is keen to be part of any groupings in the region. Symbolising this interest, the 1993 APEC meeting was held in Seattle immediately after the NAFTA agreement was approved by Congress. APEC members account for about half the world's production and 40% of trade, so it is a formidable grouping. It has a secretariat in Singapore and ten working groups covering all aspects of trade, technology and resources. Fig. 19 shows the trade flows and the considerable expansion predicted to 2001. All this must be of considerable importance to the UK if advantage is to be taken of the possibilities.

For the EU as a whole, the 12% of world trade representing EU trade outside the EU, needs to be defended and increased and represents a challenge as APEC gets underway. At the same time, APEC has to develop very considerably before its trade driven approach becomes a real possibility, let alone a reality. Anything beyond that approaching a community in the EU sense, with freedom of movement for people and finance as well as goods, is hardly on the agenda for a region which has few of the ideological and cultural ties which bind the countries of the EU together. Europe has the advantage of cohesion and should exploit that to the full.

Fig. 19 will be found overleaf.

APEC: proportion of world trade

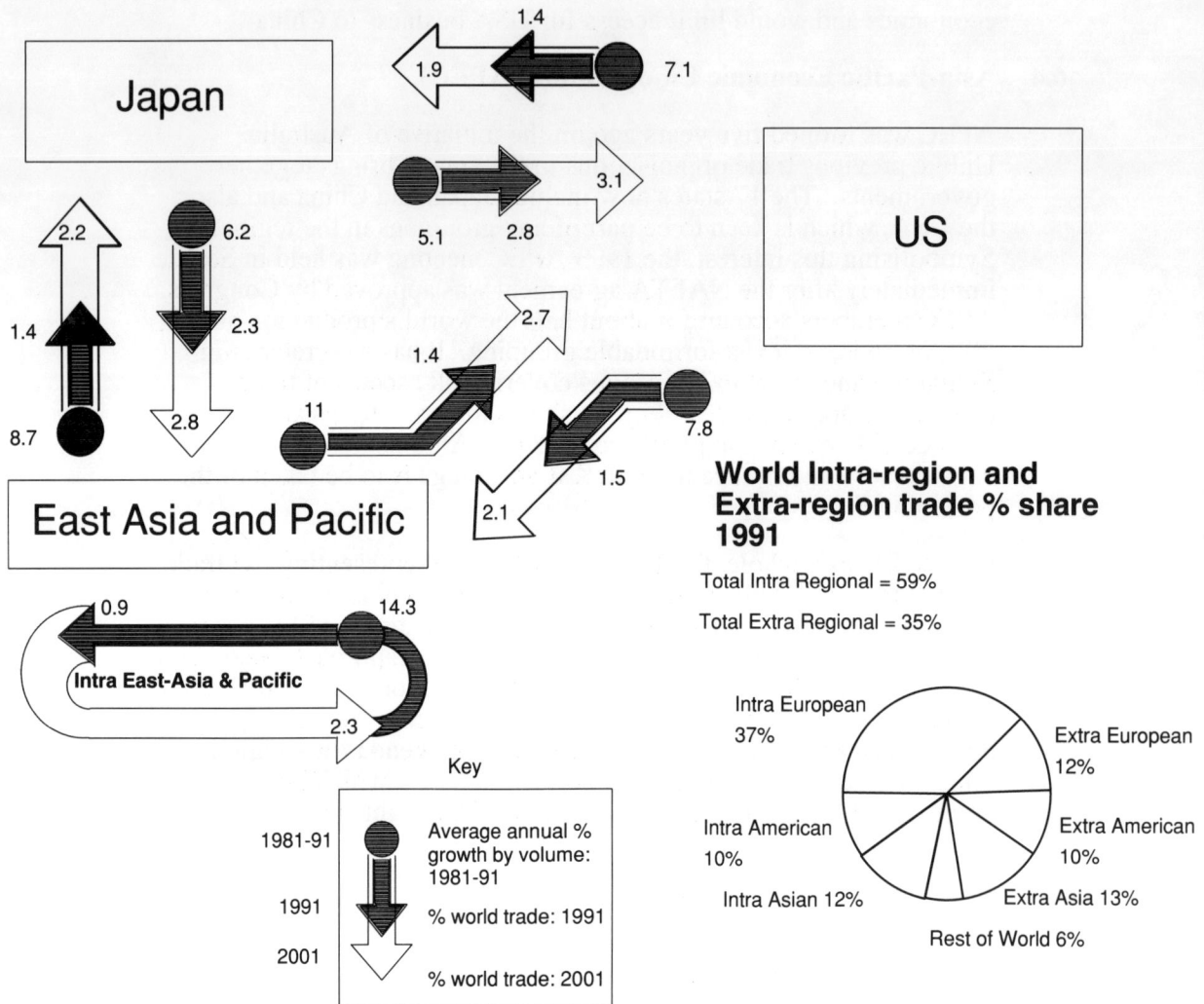

Japan

US

East Asia and Pacific

1.4
1.9
7.1
3.1
5.1
2.8
2.2
6.2
1.4
2.3
8.7
2.8
11
2.7
1.4
7.8
1.5
2.1

Intra East-Asia & Pacific
0.9
14.3
2.3

World Intra-region and Extra-region trade % share 1991

Total Intra Regional = 59%

Total Extra Regional = 35%

Intra European 37%

Extra European 12%

Intra American 10%

Extra American 10%

Intra Asian 12%

Extra Asia 13%

Rest of World 6%

Key

1981-91	Average annual % growth by volume: 1981-91
1991	% world trade: 1991
2001	% world trade: 2001

APEC economies compared

	1991 GDP ($bn)	1991 GNP per head ($)	Growth of GNP per head 1980-91 (Average annual %)	Exports 1991 ($bn)	Growth of export volume 1980-91 (Average annual %)
East Asia & Pacific	961,754	650	6.1	251,448	10.2
US	5,610,800	22,240	1.7	397,705	4.0
Japan	3,362,300	26,930	3.6	314,395	3.9
World	21,639,100	4,010	1.2	3,336,550	4.1

Figure 19 Trade Flows

Source: World Bank and GATT

PART 2

CHANGES IN THE UK

7. THE POSITION OF THE UK

7.1 UK Performance (Key Indicators)

It is clear that things are getting tougher in the world of manufacturing. To survive and prosper in this economic battlefield we must become ever more competitive with the USA, Japan and the emerging nations in Asia.

All is not gloom. We have de-regulated, privatised and adopted a market-driven, no-subsidy philosophy for our industries, ahead of our main EU competitors (Fig. 20). The resulting flexible, relatively incorrupt system could bode well for the future. Examples of successes are the record UK car production in 1993 and the position of Rolls Royce in the aero-engine industry, partly in co-operation with BMW. Public services have also had to become more accountable and efficient. This has often been painful in terms of re-structuring and unemployment, but it is fundamentally right in terms of productivity. If we had not taken this route, prosperity and employment would have suffered even more.

The UK Government should press other countries in the EU very hard to reduce their subsidies to manufacturing industry in the interests of free trade and also to maintain efforts on deregulation in the air transport field, for example, where the UK has much to gain. The EU has authorised subsidies worth about £2.3bn to state carriers since 1991 to the detriment of UK private carriers. Although an EU report has recently recommended privatisation of all state airlines, there is certain to be determined resistance on the part of Air France and Iberia, among others.

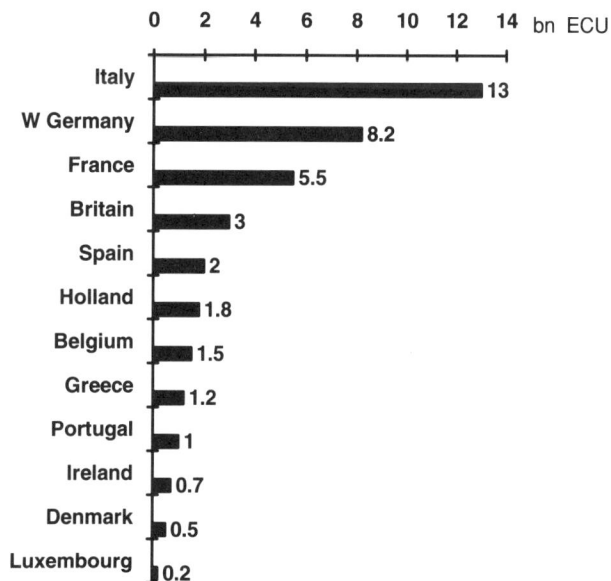

Figure 20 State Aid to Manufacturing 1990 (bn ECU)

The previous report UK-SOS of May 1992 [1], (See Annex A for the Executive Summary) focussed on four key indicators which could be

31

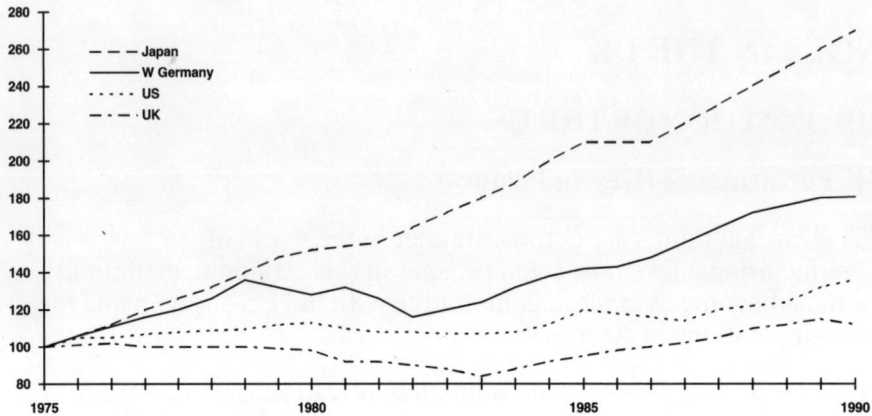

Figure 21 Manufacturing Output (1975 = 100)

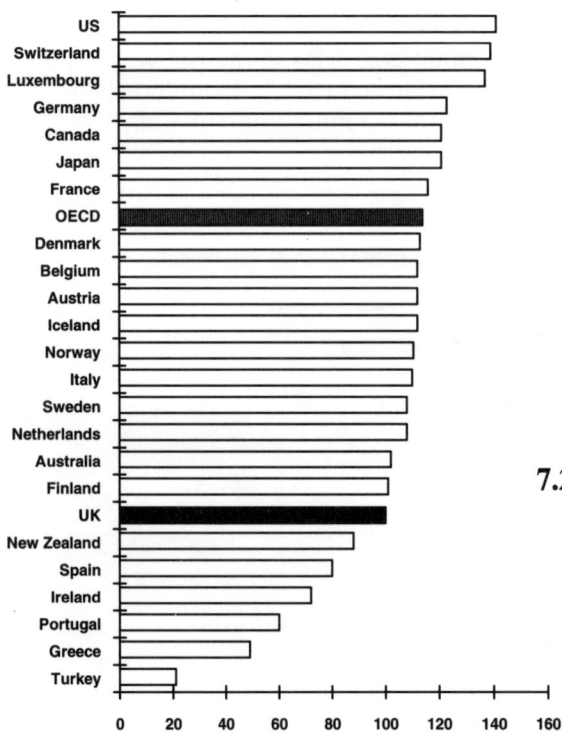

**Figure 22 GDP per Head 1991
UK = 100**

employed to analyse and assess the performance of UK manufacturing industry. They were:

*Output
Productivity
Share of World Trade
Import Penetration
and Balance of
Payments*

These can now be brought up-to-date in the light of the changing global situation and an assessment made of the need for improvement.

7.2 Output

Between 1975 and 1990 the increase of manufacturing output in the UK compared unfavourably with those of the USA, Japan and Germany (Fig 21). This is the really telling graph. As a proportion of GDP, manufacturing output fell from 32% in 1970 to just above 20% in 1992.

Fig. 22 shows that in 1991 we were only eighteenth out of the twenty-four most developed countries in terms of GDP per head.

Treasury forecasts (Fig. 23) do show a rise in GDP to 1997-98 and this is not at variance with forecasts from observers outside the UK. For example, the International Monetary Fund (IMF) has recently made the following projections for annual percentage changes in output as shown in Table 5.

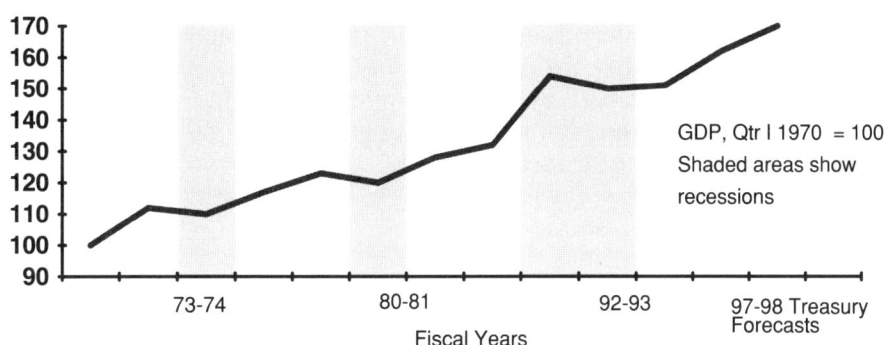

GDP, Qtr I 1970 = 100
Shaded areas show recessions

Fiscal Years

Figure 23 The Third Upswing Since 1973 (GDP)

Table 5 - Output Projections

	% annual change in output	
	1993 over 1992	1994 over 1993
World	2.2	3.2
USA	2.7	2.6
Japan	-0.1	2.0
Germany (E & W)	-1.6	1.2
France	-1.0	1.1
Italy	0.3	1.7
United Kingdom	1.8	2.8
Canada	2.6	3.8
Africa	1.6	2.6
Asia	8.7	7.1
Former USSR	-13.7	-2.4

While the star performer by this measure is Asia, it is somewhat cheering that the UK is top of the EU league and only a little behind the USA and Canada.

But these are percentage increases and in absolute terms we still have a lot of catching up to do.

More recent UK output in manufacturing, services and non-oil GDP is shown in the Treasury chart of Fig. 24 with a manufacturing upswing apparent in 1993. Fig. 25 from Oxford Economic Forecasting gives more detail with sector projections up to 1996 of annual percentage changes. These qualify for the category of low to moderate growth but will still leave us well down the league depicted in Fig. 22. We need to look for methods of accelerating growth beyond those included in the Oxford forecasts.

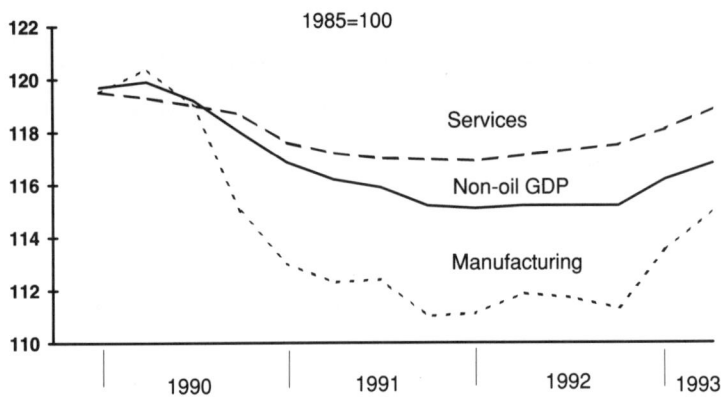

Figure 24 UK Manufacturing and Services Output and GDP

Source: Treasury

The necessary measures are listed in the "compendium of remedies" in the previous report [1] and have not changed significantly in the intervening two years. A further word of caution is necessary. It is not clear that the projections in Figs. 23 and 25 take sufficient account of the profound social and economic changes in other parts of the world described in Part 1 of this report. These changes may greatly intensify competition for trade and tend to close up markets for British goods in countries where recovery lags behind ours. There are still many hidden dangers in the present international economic situation.

7.3 Productivity

An encouraging feature in the UK manufacturing scenario is that in every year since 1981, UK productivity has shown a positive percentage growth (Fig. 26), together with steady increases in manufacturing earnings and just recently, with an actual drop in wage costs. The productivity increase has exceeded the 5% annual target set by the CBI [6] for the 1990's and the Treasury says [7] that this recent high productivity growth has contributed to the downward pressure on inflation and has translated into the lower wage costs. The manufacturing productivity trend compares favourably with developments in other G7 economies as shown in Fig. 27 but these are only percentage increases from a low starting point for the UK.

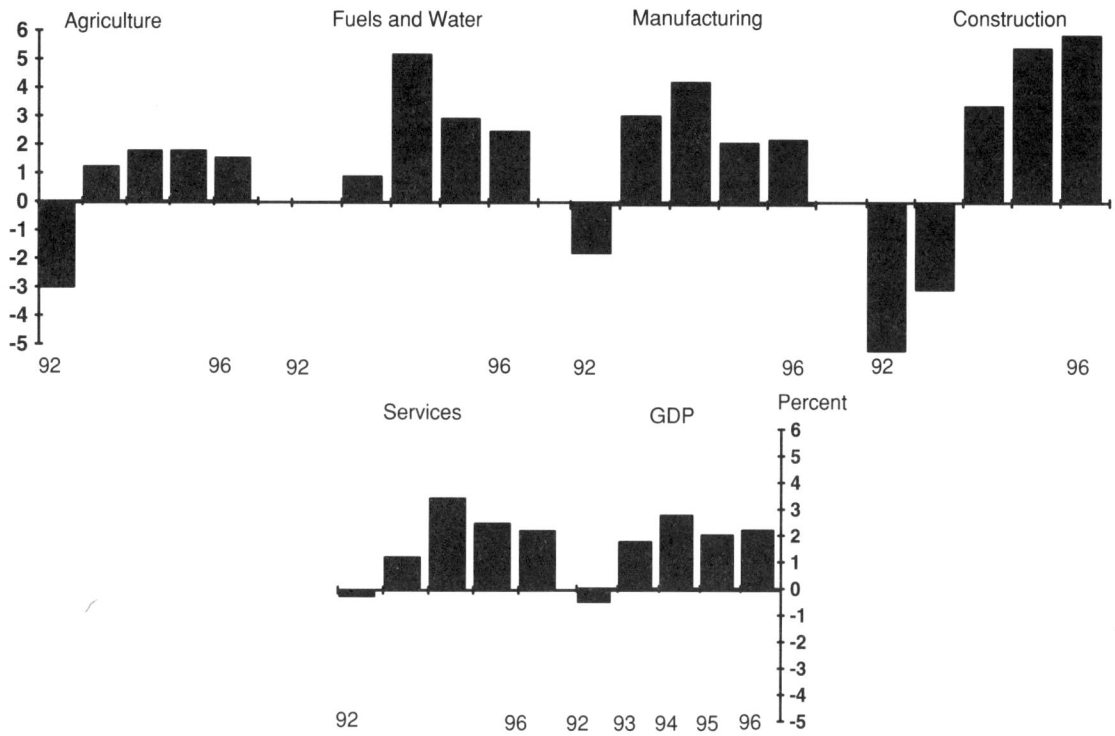

Figure 25 UK-The Forecast Recovery - Annual % Change

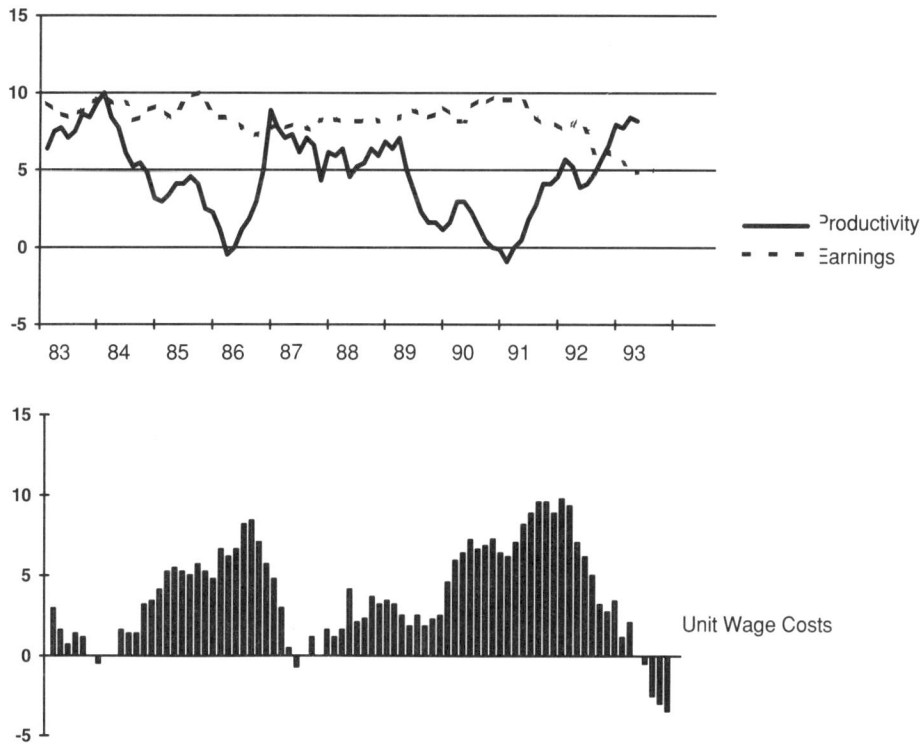

Figure 26 UK Productivity, Earnings and Unit Wage Costs in Manufacturing: percentage changes on a year earlier.

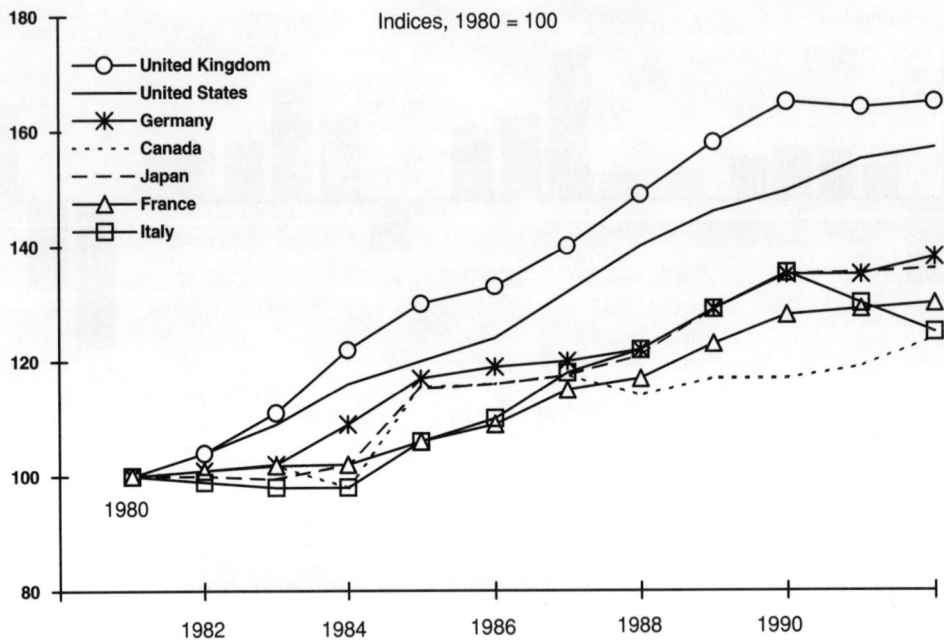

Figure 27 Manufacturing Productivity

Source: Treasury

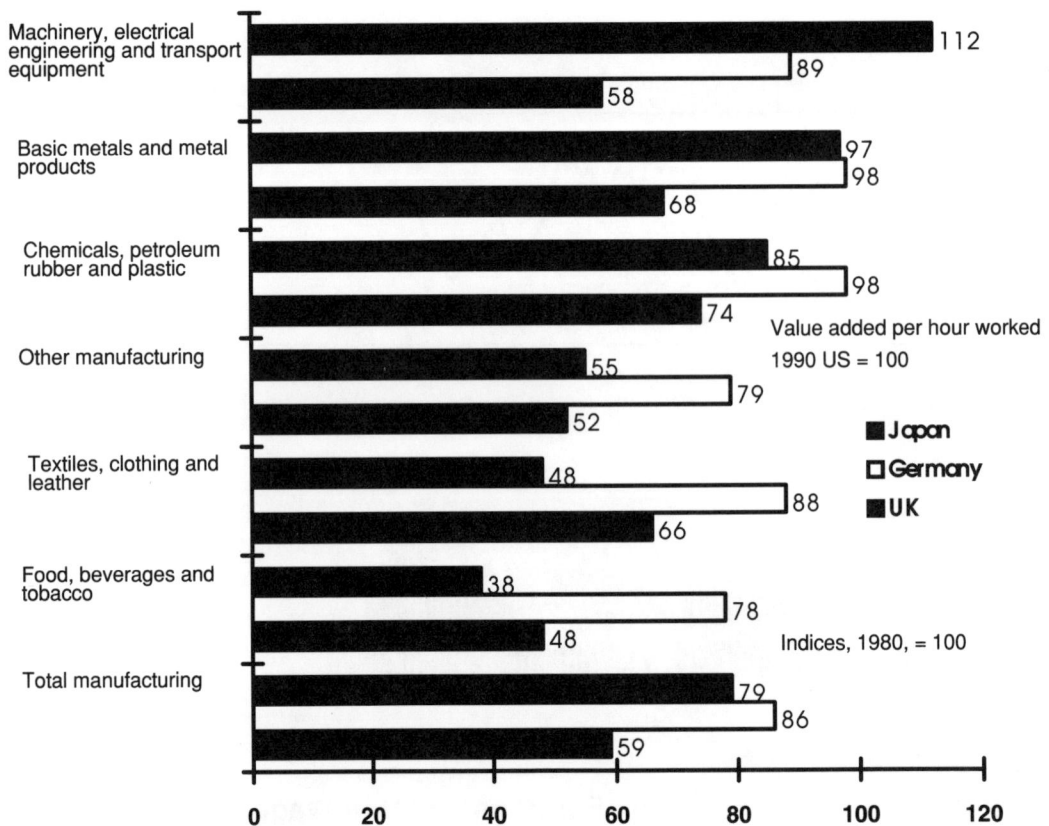

Figure 28 Productivity: Japan, Germany and UK compared with the USA

36

The DTI believes that overall we are still about 25% adrift but that the gap is closing. We therefore remain "competitive" in exports partly because of our lower wage costs.

The actual productivity positions in various industrial sectors for Japan and Germany, relative to the USA, in 1990 are shown in Fig. 28 from a recent detailed study by the management consultants McKinlay. The UK data also relative to the USA has been added for 1987 from the National Institute Economic Review [8].

Table 6, derived by the CBI from OECD data, shows the total manufacturing positions for 1991. The UK remains at the bottom and further substantial improvements are needed to maintain a long-term competitive position.

Table 6 Relative Manufacturing Productivity 1991

USA	100%
UK	57%
Japan	86%
Germany	80%
France	79%
Italy	69%

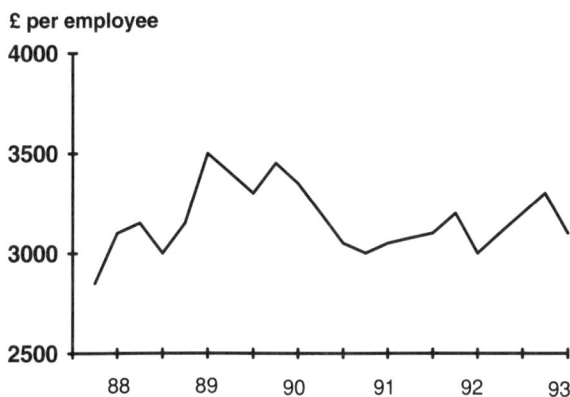

Figure 29 UK Investment remains sluggish on plant and machinery

Productivity is highly dependent upon capital investment. In the previous report[1] it was noted that during the 1980's the cumulative capital investment in UK manufacturing per employee was only 37% that of Japan and 70% of the USA and Germany. Unfortunately Fig. 29 indicates that this situation cannot have changed much since then.

Rise in Productivity	+	Flat Manufacturing	=	Drop in Manufacturing
(Figs. 26 and 27)		Output (Fig. 21)		Employment (Fig. 30)

It is the right hand side of this equation which inevitably hits the headlines. To reverse the trend we need to raise output, and hence our manufacturing base, significantly.

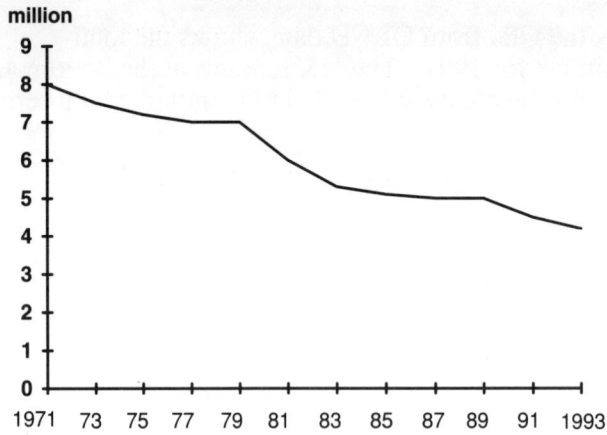

Figure 30 UK Manufacturing Employment

7.4   Share of World Trade

In UK-SOS it was shown how the UK share of world trade had diminished since 1870 and it was concluded that a reasonable share for the UK should now be around 7.8%. In the early 1970s there was a precipitous drop from about 9.5% to something like 6% and since then the figure has fluctuated around 7% with the current trend heading below this value (Fig. 31).

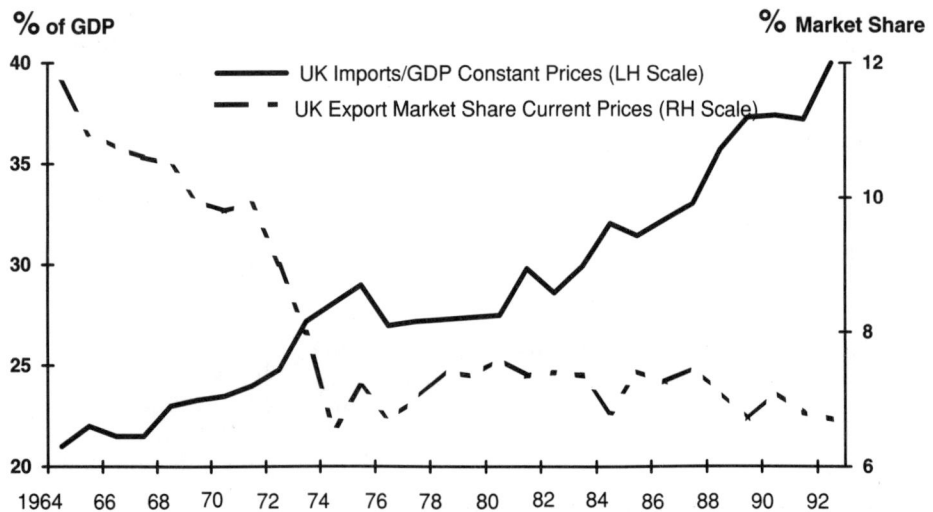

Figure 31 Trends in UK Imports and Exports

38

To stabilise or, better still, reverse this downward slope it is necessary for the UK to develop more world class manufacturing companies. A report by IBM/LBS[9] studies how our companies stand in relation to the competition and their perceptions of themselves. Among the findings are that:

- only 2% of those surveyed are truly world class but encouragingly another 42% have the practices in place to become world class.

- Britain's manufacturers are more positive than we may think - 73% believed that, either fully or mostly, they match up to the standards of their best international competitors.

- the best overall practice typically came from manufacturing sites with between 50 and 200 employees. (There is a paradox here in that the big companies with more than 500 employees are generally more productive per employee).

- the biggest inhibitor is the ability to change rapidly enough.

- leaders (the top 10%) are much more likely to use external benchmarks to improve their performance than are laggers (the bottom 10%).

7.5 Import Penetration and Balance of Payments in Manufacturing

It is often stated that the UK exports more per head than either the USA or Japan. This is true (Table 7) but is only part of the truth. We actually export less per head than other comparable EU countries and considerably less than Germany and the Netherlands. The balance against us per head is greater than for any of the other countries listed.

The destination of UK exports is given in Table 8 and it is clear that no matter what the politicians may say or do, our industry is already pan-European.

The UK trade balance (manufactured goods only) since 1970 is shown in Fig. 32. At present it could be getting better or worse. The recent UK total balances within and outside the EU, including oil and erratics, are detailed in Table 9. The worsening situation over the past three years must be a matter of serious concern.

Table 7 World Exports and Imports 1992 (Visibles)

| | Exports | | Imports | | Balance |
	% of world total	$ per head	% of world total	$ per head	$ per head
UK	5.2	3,350	5.8	3,894	-544
USA	12.1	1,788	14.4	2,208	-420
Japan	9.2	3,090	6.1	2,027	1,063
Germany	11.6	5,430	10.7	5,164	266
France	6.4	4,140	6.3	4,120	20
Italy	4.7	3,070	4.8	3,246	-176
Netherlands	3.8	9,333	3.5	8,933	400

Table 8 Destination of UK visible exports 1992

European Union (to Germany 14%, France 11%, Italy 6%)	56%
Other W Europe	8%
North America	13
Japan, Australia, NZ	4%
OPEC countries	6%
Rest of the World	13%

The Non-EU visibles deficit accounts for the bulk of our problem and has hovered around £800m a month for some time. So it is essential that our manufacturers and traders take maximum advantage of the GATT round (Section 7.6) if the UK situation is to be improved by greater exports to Non-EU countries.

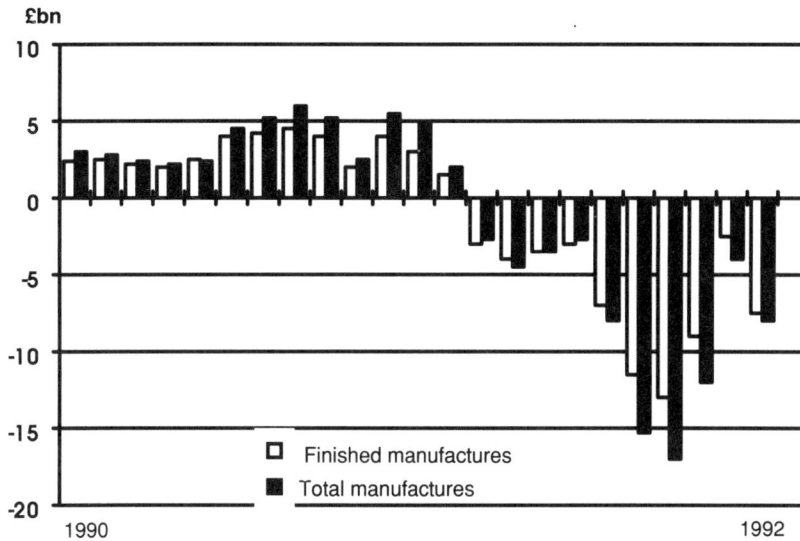

Figure 32 The Manufacturing Deficit

Table 9 UK Trading Partners £M

	Exports Visibles		Imports Visibles		Balance Visibles		Invisibles Balance	Total Balance
	EU	Non-EU	EU	Non-EU	EU	Non-EU		
1991	58,936	44,477	59,814	53,883	-878	-9,406	2,632	-7,652
1992	60,365	46,682	64,022	56,431	-3,657	-9,749	4,859	-8,547
1993	63,500	63,500	67,411	66,919	-3,911	-9,512	2,266*	-11,157

*Estimated
Source: CSO

Fig. 33 from the CBI [6] shows the sector trade balances in 1981 and 1991. Contrary to popular belief the engineering sectors have mainly positive balances but these have to pay for unfavourable balances in paper and board, food, drink, tobacco, textiles and clothing. Whatever improvements are needed in manufacturing, those sectors also need to perform much better by way of import substitution and export promotion.

Electrical and Electronic Engineering shows a slight balance of payments deficit unlike chemicals, aerospace and mechanical engineering which still export considerably more than they import.

£ bn at 1991 Prices

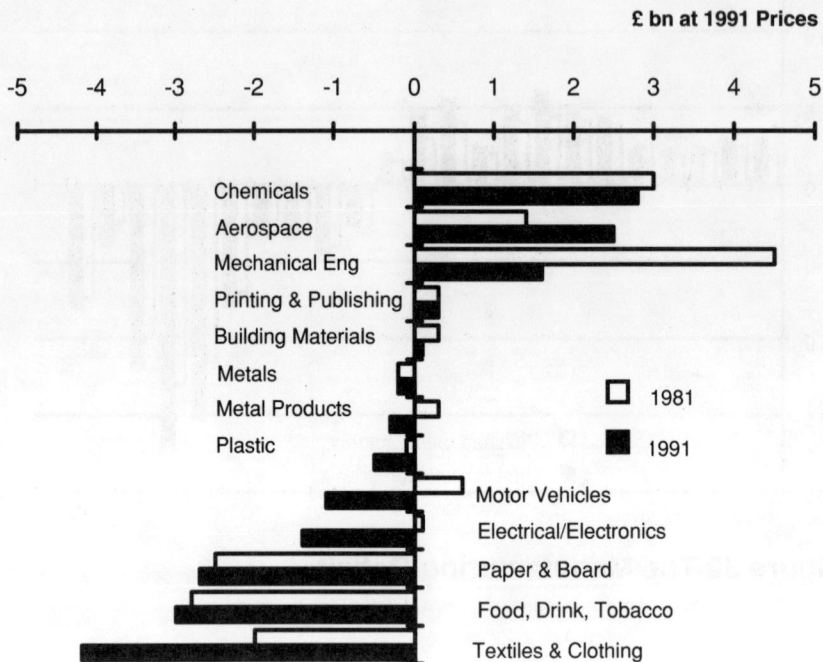

Figure 33 Sector Trade Balances

Source: CSO

Electronics is a key area, so its negative balance is of considerable concern. Nevertheless, the UK has the fourth largest electrical and electronics industry in the world after the USA, Germany and Japan. It is now the second largest industry in the UK, employing some 330,000 people and investing £1bn a year in capital and £2bn in R & D. The UK has developed particular strengths in telecommunication systems, data processing and software, but weaknesses need to be remedied in market penetration into Europe and in the components sector in order to rectify the trade balance.

7.6 The GATT Round

The successful completion of the Uruguay round of the General Agreement on Tariffs and Trade (GATT) in December 1993 marks a significant international movement towards free trade among 117 countries (though not yet China) and comes into force in 1995.

The new agreement covers an impressive range of goods and services, $4,500bn compared with the previous Tokyo round of only $300bn.

Benefits due to the reduction of protectionism build up to an estimated $274bn of extra annual trade by 2002, distributed as shown in Table 10.

Table 10

	Projected Annual Benefit of GATT $bn
Australia and New Zealand	2
Canada	7
EU	71
EFTA	38
Japan	42
USA	28
Rest of World	86
	$274bn

Most of the savings are associated with the agricultural sector but the agreement also covers tariffs on manufactured goods, and intellectual property rights. The annual savings will constitute about 1% of world trade. Their significance is great because of the huge range of goods to which the GATT applies, though this is not as wide as the USA would have liked. Subsequent negotiations on increasing the scope appear to have failed.

The figures indicate that the EU could be a substantial gainer from the GATT round. The main British beneficiaries are expected to be in the areas of consumer goods and pharmaceuticals. However, losers are predicted to be in many of our traditional areas such as bulk chemicals, general manufacturing, mineral extraction, electrical and electronic manufacturing, aerospace and textiles.

If, as seems likely, this GATT round will accelerate the shift of production to low labour cost economies, then the continued decline of UK manufacturing could result in further depredations on our balance of payments which would be difficult to contain. For example the negative balance in textiles shown in Fig. 33 could grow even worse while the present positive balances due to the engineering sectors would reduce. The net effect could be very serious indeed. Part of the answer has to lie in the development of "agile" factories built here to balance the advantages of cheap labour elsewhere. The chain of reasoning is given in Conclusions, Section 9.

The GATT gains will go to those countries geared to take full advantage of the opportunities and to attract inward investment. Early estimates indicate that 400,000 new jobs <u>could</u> be created in British industry. But success will depend upon the positive signal on free trade from world leaders being acted upon swiftly by our business community. The improvements in exports will need to be directed at countries outside the EU because within it tariff barriers have already been abolished. The USA, Australia and Asian countries suggest themselves as the main targets because import taxes on many product taxes will be cut from their present 40% levels.

8. CHANGES IN UK DOMESTIC FACTORS

8.1 Inputs affecting UK Performance

There are a considerable number of domestic factors which influence the four key indicators described in Section 7, and so affect manufacturing activity and competitiveness.

The 1992 UK-SOS report[1] identified six major areas which, for convenience, were termed "inputs" because they are largely within UK control and feed directly into industrial performance. They are:

> *Education and Training*
> *Research and Development*
> *Innovation*
> *Management Style and Quality*
> *Size of Company*
> *Finance, Investment and Government Policy*

There has been a great deal of development in most of these areas over the past eighteen months reflecting, in a way, the rapid shifts in the world scene.

8.2 Education and Training

Considerable changes have taken place in the education and training systems, from primary school to postgraduate level. Collectively the changes represent a radical departure against the background of the Government's twin aims of improving student achievement and gaining better value for money.

There is still a long way to go in education and training, as shown in the latest figures from the long-term researches of the NIESR [10] (Table 11).

Table 11

	Proportion of each age group reaching equivalent of GCSE Grades A-C in Maths, Science and National language 1990/1	Proportion of 16-19 age group in full time education or part-time education and training1991
Germany	62%	79%
France	66%	76%
Japan	50%*	94%**
England	27%	56%

* 1986
**15-18 year olds

Comprehensive proposals for change from nursery school to university were put forward in November 1993 by the National Commission on Education established by Sir Claus Moser [11] and just recently, in response, the Government has committed itself to introduce, at some time in the future, universal nursery education, probably on a voucher system.

8.2.1 Schools - National Curriculum

The National Curriculum has had to be considerably modified as the price for teacher co-operation but is now proceeding more smoothly after recent changes on the treatment of tests for 7 and 14 year olds. The subject of technology is still contentious. Even the revised curriculum gave the appearance of being too complicated, ambitious and, in some ways, ambiguous for the average teacher of the subject to handle. However, the very latest proposals (September 1993) from a committee reporting to Sir Ron Dearing are much more straightforward and are structured around readily understood themes. A big improvement, particularly as they seem likely to be accepted and implemented without much rancour. The "Good Technology Guide", just issued by the Department for Education and based upon numerous inputs from practitioners and educational organisations, should prove to be a key resource in developing the teaching of National Curriculum technology.

To assist the teaching of technology The Engineering Council has recently launched the "Technology Enhancement Programme" sponsored by the Gatsby Charitable Foundation, and has given £10,000 each to 22 schools. 62 schools belong to the programme.

8.2.2 Schools - A levels

It is an encouraging sign that the number of 16 year olds staying on to take A-levels is steadily increasing, although the full-time participation rate still lags behind our competitors. In 1993 candidates took more than 750,000 subjects - a record - and achieved a record overall pass-rate of 81% (grades A-E) with 48% in grades A to C. A disturbing feature, however, is that Maths, Physics and Chemistry are losing out to social science subjects. In 1993 Maths accounted for only 8.9% of subject entries and physics a mere 5.1% - see Table 12.

Table 12

	A level entries in Maths and Physics	
	1989	1993
Maths	82,000	64,000
Physics	44,000	37,000

The Government is now seriously disturbed about this trend and, among other measures, has made proposals for a national network of self-governing technology colleges. Developed from existing secondary schools, these will operate a syllabus in line with the National Curriculum, but with a special emphasis on technology, science and mathematics. Industry is also expressing deep concern about future mathematics and science teaching and making suggestions for educational improvements.

There is a large and growing[11] body of opinion which would like to see a fundamental revision of "A" levels to a broader base of five subjects on the lines of the "Higginson" report several years ago. This is being resisted by the Government on the basis of defending the "gold-standard" represented by the "A" levels but also, it is to be suspected, because of the pressure which would be applied for 3-year degrees to be extended to 4-years at very considerable expense.

It may be that at least part of the proposed "Higginson" reforms will be achieved by the "back door" as access widens to include GNVQs (see below) and modular courses become the norm.

8.2.3 National Vocational Qualifications (NVQ)

The NVQ system continues to develop - albeit more slowly than the Government hoped. Also the new general qualification, which is based upon a combination of school lessons and work experience, (GNVQ) has now been introduced. Many universities are considering how this new qualification, which is based upon a combination of school lessons and work experience, can count towards entry into undergraduate courses. Its different approach to the traditional knowledge-based "A" levels involves problems of comparability, not all of which are easy to solve. In fact a great deal of nonsense is talked about the contrast between "vocational" and "academic" courses. In engineering, each course should surely combine theoretical and practical aspects to the appropriate level.

The debate continues about the soundness of the NVQ approach and it is clear that much more work needs to be done by the controlling body, the National Council for Vocational Qualifications (NCVQ) to convince educationalists and employers of the soundness of the "competence based" approach. In particular, even if NVQs demonstrate competence in an immediate task, will they, in the long run, produce a work force with a broad range of flexible and competitive skills? This question becomes sharper, the higher up the educational ladder. Certainly at degree level the whole emphasis is on fundamental knowledge and understanding. Other questions involve monitoring and in-house testing which some critics feel leads to variable standards. City and Guilds (C&G), one of the leading organisations running NVQ recognised courses has recently expressed considerable concern about the costs of the new system compared with the old where students were taught and examined in the traditional way. The increased costs fall on employers too and this is an inhibiting factor causing employer resistance. Some employers fear that workers may expect more pay for attaining NVQs!

To stimulate the effort, a business-led advisory group has recently been established by the Government (National Advisory Group for Education and Training Targets). As a target, the CBI wants 80% of young people to have achieved NVQ level 2 by 1997. The figure is now about 50%.

There is now much better collaboration between the Business and Technician Education Council BTEC and NCVQ which augurs well for the future. However, a far reaching rationalisation of the whole system would be helpful to young people, employers and education and trainers alike, besides the public at large. At present it must be profoundly confusing to most people. Who but the experts, know the difference between BTEC and TEC? We keep adding elements to the system without removing others, so that we now have a plethora of qualifications and organisations: GCSEs, "O" levels (for overseas candidates), A-levels, A/S-levels, BTEC, NVQ, GNVQ, C&G and Royal Society of Arts (RSA), even leaving aside the Scottish variations.

There is an immediate need for clearer statements to the public about the standards, aims and routes relating to qualifications. The subsequent task must surely be some drastic simplification of this jungle to eliminate duplication and to make the whole system more coherent and accessible. This simplification must also extend to the Government's own lines of responsibility which sometimes give the impression of opacity even if the arrangements are administratively convenient. For example, the Department of Employment is responsible for NVQs but the Department for Education for GNVQs.

8.2.4 Training and the TECs

The evidence is that employers are generally becoming more convinced of the importance of a skilled and flexible workforce though many are ever constrained by finance in their efforts to improve. The National Training Awards have a good mix of entries from small, medium and large companies and the hundreds of winners in 1992 and 1993, using the NTA logo, reflect a broad spectrum of industry and commerce.

The Training and Enterprise Councils (TECs) were launched 3 years ago and there are now 82 of them. The Boards are composed mainly of business executives and the total funding amounts to some £2.3bn annually.

The original brief of the TECs was to run Government programmes for the unemployed more efficiently, address local skills needs and set priorities for economic development with other agencies. In the event, they have been forced to concentrate on the programmes for youth and adult unemployment and the "enterprise" part of their mission. Enhancing the skills of the existing work force and developing business locally has had to be relatively neglected. Many of the boards now want to redress the balance to help meet the needs of a competitive economy, but to do so will require more flexibility from Government and better co-ordination among the TECs themselves. The general consensus is that performance of the TECs is patchy with some good examples, particularly in the Midlands and the North, but with quite a number of indifferent ones. Different perspectives of the TECs give rather different pictures:

- a Financial Times survey of the TECs in May 1993 found that most TEC directors were satisfied with progress generally and particularly in the relationships established with schools. Almost all saw youth training at 16-19 years of age as an important area. More than 80% welcomed the Government scheme under which employers are given a Government subsidy to take on someone who has been without work for a long time.

- directors of TECs called for greater co-ordination of Government departments involved in education, training and enterprise. They also want a structured partnership among the several bodies responsible for developing local economies.

- in January 1994 a searching report by the London School of Economics [12] although sympathetic in tone, was critical of many aspects of the TECs. It says "Far from fulfilling their mission of being world-class business-led entities, supporting mainstream development of skills and enterprise, the TECs are likely to drift into a second tier role as business leaders lose interest and staff with few business skills exert even greater control. Business empowerment has become a sham".

48

- it seems that in Scotland the equivalent Local Enterprise Councils (LECs) have had greater success because of an arms-length relationship with the Government. The LSE study concludes that a rescue package is needed urgently - and, one may add, a package which ensures that the funding matches the tasks given to the TECs.

Part of the solution lies in the November 1993 report from the CBI "Making Labour Markets Work" which puts forward more than 50 proposals to promote the enterprise part of the TECs mission and to reduce bureaucracy about which there is growing concern.

In December 1993 the Government announced a new apprenticeship scheme to boost training in technical and supervisory areas. Details are still awaited but it seems that the scheme will be run using the existing "youth credits" under which vouchers for buying training are made available to young people through TECs.

In the spring of 1993 the total number of apprentices was 245,000, down 80,000 on the previous year. The new scheme is planned to add 150,000 apprenticeship places, but not necessarily of the traditional type. It is too early to judge whether the new scheme will be of benefit or whether, as some fear, it will prove to be simply a revamp of an existing unsatisfactory system.

8.2.5 Universities

Since the 1992 UK-SOS report[1] the most important development in higher education has been the transition of the polytechnics into universities. The "binary-line" has been abolished and we now have a unified system of higher education with approximately 100 Universities or University Colleges all funded through a single body, the Higher Education Funding Council for England (HEFCE) which has replaced both the University Grants Committee (UGC) and the Polytechnic and College Funding Council (PCFC). The same body also funds 77 other colleges in England.

Consequential changes have been the establishment of a unified Committee of Vice-Chancellors and Principals (CVCP) and a unified University entrance system (following the amalgamation of UCCA and PCAS).

The underlying philosophy has been a liberal one - to enlarge access to higher education for suitably qualified students and provide an environment of excellence, not just in research but also in teaching. This has worked so well that it has sometimes caused embarrassment to the planners because increased student numbers carry the commitment of increased expenditure. Consequently the Government has taken measures to rein back the potential rate of increase in 1994.

Even so, the present and projected increases in the numbers of undergraduates is producing a situation where the long-standing system of Government grants to students for fees and maintenance is becoming unsustainable if expansion is to continue and the quality of teaching is to be maintained.

A new system based on loans is necessary but, as yet, there is no agreement on what form it should take. At present, no body wants to be responsible for the decision; the Government, nor the CVCP, but a solution is needed quickly if the stability of the system is to be preserved and if students from poorer homes are not to be grossly disadvantaged. The National Commission on Education [11] has proposed a loan system with repayments by graduates through the tax system similar to Australian practice, but the Government's policy seems to be developing along the lines of gradually replacing maintenance grants and possibly fee support by low cost loans which are repaid through the banking system.

Further consequences of the recent changes in higher education are:

- the new monolithic nature of the system means that potentially the CVCP is a much more powerful body than it was in the old system with the binary divide. In those days the Government often played off the polytechnics against the universities on such matters as expansion and unit costs. That will no longer be possible if the CVCP can get its act together and speak with a strong single voice. At present it seems unwilling to come to a consensus on important fundamental issues involving resources for teaching and research and is not as effective as it otherwise could be.

- the pressure on resources, most obvious in the employment of fewer academic staff for more students, is leading to a mismatch between the amount universities can teach in the time available and the requirements of professional bodies to satisfy their criteria for competence to practise (see 8.2.6). It is possible that some university departments will become "bankrupt" and that low grade universities will be forced to close or amalgamate.

- the publication of "league tables" of universities on the basis of research, facilities, completion rates, value-added by good teaching etc., could gradually influence student choice as each university strives to be at the top of a particular "league" of excellence.

- the increasing number of graduates in subjects other than the "hard" sciences and engineering has led to more graduate unemployment. Graduates and employers alike will need to re-orientate their views on employment to raise general standards throughout business, industry and commerce. This will require that graduates are more flexible in the kind of work they take and that employers realise the possibilities of a better educated group at deeper levels in their organisation.

8.2.6 Engineering Degree Courses

In spite of the greater numbers of students entering higher education it is still difficult to fill the available places in engineering degree courses with well qualified candidates. This is certainly attributable (in part) to deficiencies in mathematics and physics teaching in schools (Table 12). Numerous attempts have been made by the

Engineering Council to rectify this situation through, for example, its "Neighbourhood Engineers Scheme", linking engineers with their local schools and its Manufacturing Engineering initiative which has improved student numbers, equipment and industrial linkages.

The latest move (summer 1993) is the "Top Flight" bursary scheme which will provide suitably qualified students (AAB at A-level) with an additional £500 per year maintenance grant, for the duration of their engineering degree courses. The laudable aim is to make engineering a more attractive educational and career goal for bright young people. It has to be said, however, that a similar scheme, piloted about a decade ago, failed to attract significant numbers of extra candidates. The effect was simply to give more money to some of those who were going to take engineering degrees anyway. All the same, in the present economic climate this may be a worthwhile objective, so that well qualified students can be relieved of financial worry and so concentrate on their engineering studies.

In engineering, as in other professional subjects, the general squeeze on university resources is causing a mismatch between what is possible in a 3-year degree course and the requirements of the professional bodies. The signs are that this problem is about to affect the engineering profession acutely.

The standards of qualification are defined in the Engineering Councils publication "Standards and Routes to Registration" known as SARTOR. This has the general backing of the Engineering Institutions and of industry. Its operation, depends upon degree courses in universities being able to meet the required academic standards. This is checked by a system of course accreditation operated by the engineering institutions. It is becoming questionable whether the academic standards for potential Chartered Engineers can be achieved in a 3-year degree course with the reduced resources available. Already the Institution of Civil Engineers has withdrawn accreditation from at least one university degree course and if this happened on a large scale, the standards and basis of entry to the profession could be jeopardised.

A recent Engineering Council discussion paper [13] explores this problem and suggests that the way forward may be to line up the Chartered Engineer accredited standard with an extra year of study for fewer students - the MEng rather than the BEng degree. The discussion document also commendably proposes more specific criteria than hitherto for management and business studies and for a foreign language in relation to all three grades of registration - Chartered Engineer, Incorporated Engineer and Engineering Technician. The inclusion of these additional studies is to be welcomed provided that the resources are available for teaching and assessment of competence.

51

The Engineering Council discussion paper was written against the background of a research report from the Policy Studies Institute [14] which concluded that:

- engineering skills requirements will grow at the same rate as for professional qualifications as a whole, that is, in the range 1.7 to 2.1% per annum.

- a challenge for the education and training system will be not only to provide qualified people in sufficient numbers, but to develop skill content in such a way as to respond to the varied changes in engineering work.

- education and training provision should take into account that engineering skills are important across the whole economy in high tech, middle-tech and low-tech industries alike and in production and services alike.

The conclusions are also reflected in a recent lecture [14] given by Howard Davies, Director General of the CBI.

8.3   Research and Development

8.3.1 The White Paper

Since the previous UK-SOS report[1] the Government has published its first White Paper on Science for 20 years[16]. "Realising our Potential - A Strategy for Science Engineering and Technology" redefines much of the Government's involvement in the scientific and technical arena. It explicitly recognises that science, technology and engineering are intimately linked across the whole range of human endeavour; educational, intellectual, medical, environmental, social, economic and cultural.

Among the main proposals contained in the White Paper are:

- the promotion of awareness of the importance of innovation among senior managers. Measures include support for industrialists as visiting professors.

- encouragement to Business Schools to develop a national modular Masters degree in the management of technology. (Should this only be available, with Government support, to those with an engineering or scientific qualification?)

- Technology Foresight. The purpose is to gain early notice of emerging key technologies and to assess scientific and technological developments which could have a strong impact on industrial competitiveness. A high-powered steering group has been established under the Government's Chief Scientific Adviser and expert panels are being formed.

This is a valuable development which should help the targeting of R & D money. There was initial concern by the CBI that the approach was too narrow in terms of the initial list of nine technology areas and that there was too much concentration on technology push rather than market pull but later announcements rectify this impression.

Technology Foresight must be a long term programme and tenacity is needed. The IEE has suggested a pilot scheme in one selected area to test the credibility of the process and establish a track record to instil confidence in the programme as a whole.

To carry the process forward, panels of industrialists and academics are being established for each of 15 sectors with the aim of initial reporting early in 1995.

- from 1994 the Government will publish each April a "Forward-Look" giving a longer term assessment of the portfolio of publicly-funded work to set perspectives over a 10-year period - a useful prospect.

- a new Council for Science and Technology (CST) will advise on the balance and direction of research and will be chaired by the Chancellor of the Duchy of Lancaster with largely independent members, but including the Chief Scientists of Government departments.

- candidates for a PhD degree must first take a Masters degree course. This proposal has had a mixed reception. It is not yet clear whether the Masters programme would be the first year of a 3-year PhD course or whether the total would be four years. Either way it may entail difficult finance and logistics problems for universities. The Office of Science and Technology has now issued a consultation paper on the future of the PhD.

- other important changes affecting the research councils and universities are also included in the White Paper and are covered in 8.3.3.

8.3.2 Government Research and Development [17]

The total picture for UK Research and Development is shown in Fig. 34. Of the total spend of some £12bn, the UK Government contributes £4.2bn Non-Government UK sources provide £6.4bn and the remainder comes from overseas.

The Government spends £1.6bn while £2.1bn goes to Higher Education mainly to Research Councils and for the "dual support" system to Universities. Fig. 35 shows the latest available time sequence 1985-91 in real terms. Disappointingly in 1991 the total GERD (Gross Domestic Expenditure on R & D) was **down** 7%, a comparatively large reduction which gives cause for concern. The Government's own expenditure decreased, even more, by 10.5% and further reductions are planned up to 1995-96.

Total GERD in 1991 - £11.9 bn

Sectors providing the funds £ bn

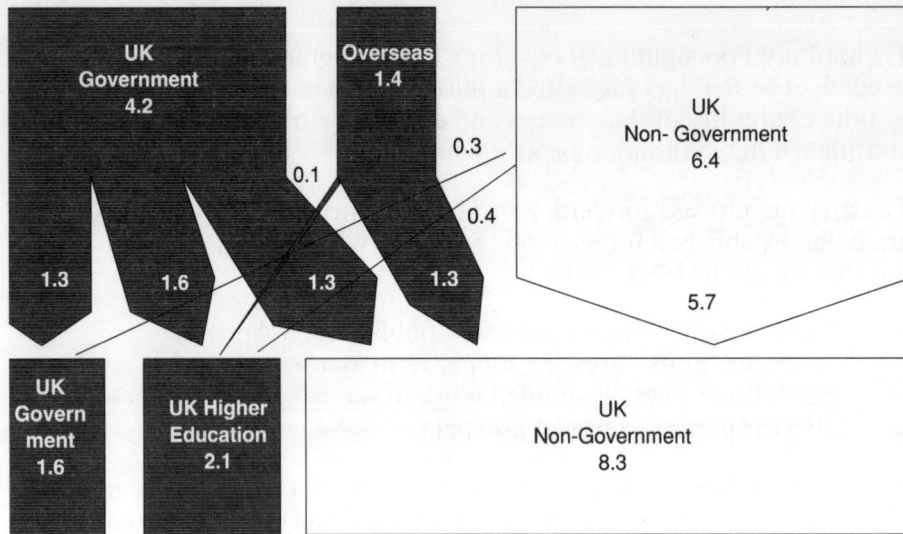

UK Government 4.2

Overseas 1.4

UK Non-Government 6.4

0.1

0.3

0.4

1.3   1.6   1.3   1.3

5.7

UK Government 1.6

UK Higher Education 2.1

UK Non-Government 8.3

Sectors carrying out the work £ bn

Figure 34 Total R & D in the UK (GERD)

£ million at constant prices

1985
244
70
918
4963
5410
11608

1986
242
74
1111
4921
5835
12183

1987
257
86
1106
4794
6116
12359

1988
267
93
1150
4499
6546
12556

1989
292
93
1307
4644
6668
13004

1990
330
90
1487
4506
6407
12820

1991
362
90
1397
4705
5981
11906

☐ Private non-profit
■ Higher Education
■ Overseas
☐ Government
■ Business
■ Total

Figure 35 R & D expenditure in the UK by funder 1985-1991

54

The distribution of Government civil and defence R & D is shown in Fig. 36. Basic research dominates under the civil heading and experimental development under the defence heading.

The distribution by socio-economic objective appears in Fig. 37. At less than 8% of the whole Government funding of R & D, industrial development seems very low.

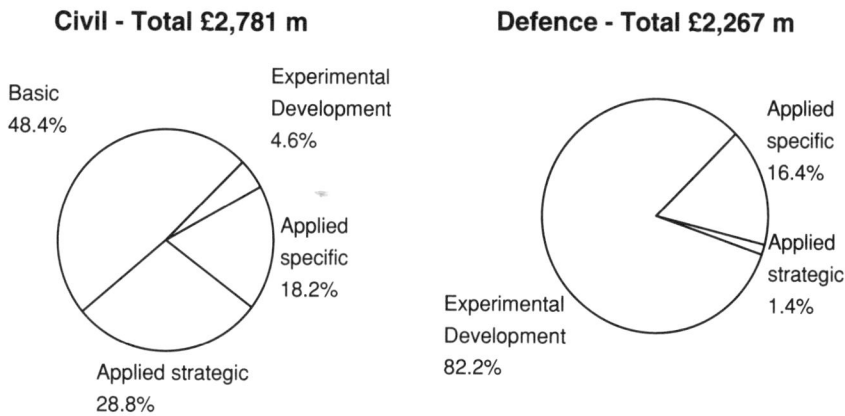

Civil - Total £2,781 m　　　　　**Defence - Total £2,267 m**

Basic
48.4%

Experimental
Development
4.6%

Applied
specific
18.2%

Applied strategic
28.8%

Applied
specific
16.4%

Applied
strategic
1.4%

Experimental
Development
82.2%

Figure 36 Distribution of Government R & D Expenditure 1991-92

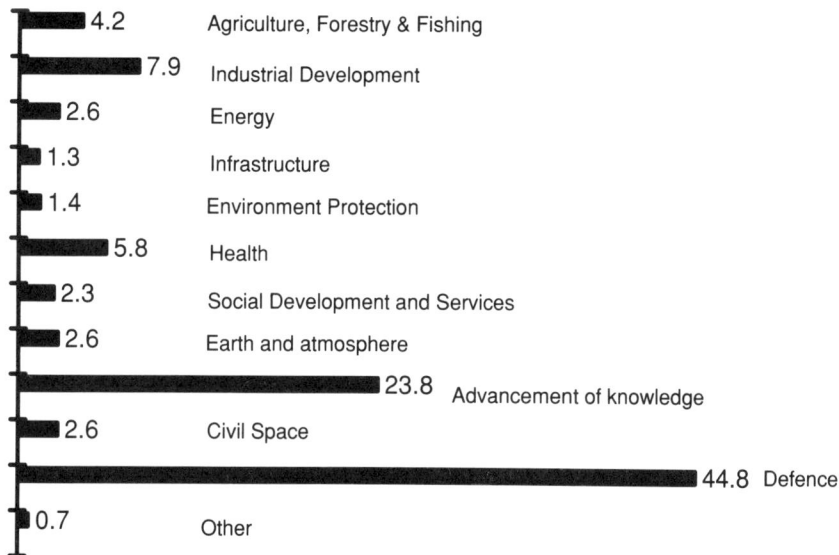

4.2   Agriculture, Forestry & Fishing
7.9   Industrial Development
2.6   Energy
1.3   Infrastructure
1.4   Environment Protection
5.8   Health
2.3   Social Development and Services
2.6   Earth and atmosphere
23.8   Advancement of knowledge
2.6   Civil Space
44.8   Defence
0.7   Other

Figure 37 Distribution of Government Funding of R & D 1991- 92 by Socio Economic Objective

Table 13 Government funded R & D as % of GDP, 1991

	Civil	Defence	Govt Total	Overall Total
UK	0.49	0.40	0.89	2.08
France	0.89	0.53	1.42	2.42
Germany	0.87	0.11	0.98	2.58
Italy	0.70	0.06	0.76	1.38
Japan	0.42	0.03	0.45	2.86
USA	0.48	0.71	1.19	2.78

International comparisons from six principal OECD countries of Governmental expenditure on R & D in 1991 are given in Table 13.

The UK and the USA have much the highest ratio of defence to civil expenditure and a principal feature of this table is the very high amount of non-governmental R & D in Germany and Japan, neither of which have significant defence expenditure. Also, as their economies are much larger than ours, their actual civil expenditure is very much larger (see also Table 14).

8.3.3 The Research Councils and the Universities

The White Paper "Realising Our Potential" besides guiding the direction of research also introduced a major re-organisation of the Research Councils. In future, the White Paper states, "decisions for priorities for support should be much more clearly related to meeting the country's needs and enhancing the wealth creating capacity of the country."

The new administrative and committee structure for the Research Councils reflects this priority and represents a major change in philosophy. Since universities are a main beneficiary of Research Council funds, they too will be significantly affected by the change in emphasis. What it means is that in future universities will, in effect, be the contractor carrying out research which can be demonstrated to be industrially and commercially useful.

The biggest change in the structure is of great consequence to engineering and manufacturing. Instead of the old Science and Engineering Council (SERC) the new arrangements involve two separate Research Councils. At long last we have a Research Council for Engineering and Physical Sciences with the mission:

> To promote and support high quality basic strategic and applied research and related postgraduate training in Engineering, Chemistry, Physics and Mathematics, placing special emphasis on meeting the needs of the users of its research and training outputs, thereby enhancing the United Kingdom's industrial competitiveness and quality of life.

There will also be a Particle Physics and Astronomy Research Council to look after the SERC's "big science" responsibilities. While the new Engineering and Physical Sciences Research Council has generally been welcomed by academic and industrial engineering interests, the House of Lords Select Committee on Science and Technology has expressed fears that basic science and the training of PhD students may suffer and that topics on the borderline between the new Research Councils could get left out.

The shake-up of the Research Council structure alone would have had far reaching effects on university research workers and teams but there are also other major factors which affect the spectrum of university research. Already some University Grants Committee (now HEFCE) funding has been diverted to the Research Councils so weakening the traditional "dual-support" system by means of which research as well as teaching was supported on a blanket basis in the "old" universities. In spite of a Government commitment in the White Paper to maintain dual support it is likely to be further weakened so that the "new" as well as the "old" universities can compete for the available research funds by applying to the Research Councils.

Furthermore, the research quality exercises which have been undertaken in the university sector over the past few years have been employed to link the distribution of the research element of the dual-support finance to research quality ratings. Only those few universities with uniformly high ratings will get research funds across-the-board from the Higher Education Funding Council in future. Inevitably this will lead to a situation where there are perhaps 15 research orientated universities with the other 85 or so receiving research support in only a few limited areas. While this may be healthy from the point of view of value-for-money in research, it is inevitably going to lead to questions about the quality of first and higher degrees in the "non-research" universities. (See also 8.2.5 on the funding of universities generally).

8.3.4 Industrial spending on R & D

Non-Governmental funding for R & D amounted to £6.4bn in 1991 (Fig. 34) and figures from 1985 are shown in Fig. 35. It is sad to see that the business contribution in 1991 dipped by 10.3 % from its peak of 1989.

Table 14 shows the international economic situation with respect to R & D, for six major countries.

Table 14 OECD Science and Technology Indicators 1991

	GDP £bn	Total GDP Growth 1985-91%	Domestic Product of Industry £bn	Gross Expenditure on R & D £bn	Total GERD ** Growth 1985-91%
UK	572	8.2	411	11.9	6.6
France	660	9.3	503	16.0	10.6
Germany	857	11.2	628	22.1	10.3
Italy	620	9.3	494	8.5	13.1
Japan	1,503	11.2	1,191*	42.9	13.0
USA	3,524	8.4	1,889*	98.0	7.5

*   1990
** Gross Expenditure on R & D

It emerges that between 1985 and 1991 we had the smallest GDP growth rate and the smallest R & D growth rate.

The sectoral distributions of R & D performed in manufacturing industry in the years 1985, 1990 and 1991 are shown in Fig. 38 at 1985 prices. With the notable exception of chemicals, no sector has increased and some have dropped sharply.

Fig. 39 brings out in a remarkable way the contrast over the past quarter century between R & D in pharmaceuticals and that in manufacturing industry generally.

In the UK-SOS report[1] there were clear indications that for smaller companies, especially those in relatively low-tech areas, less was spent on R & D than in equivalent enterprises abroad. Has this situation changed?

1985
Total 4,673 m

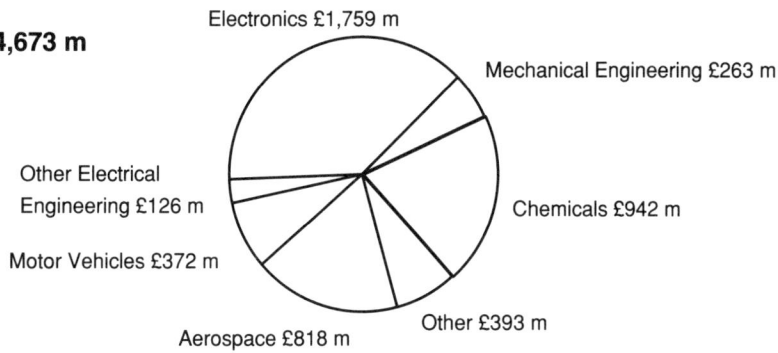

Electronics £1,759 m

Mechanical Engineering £263 m

Other Electrical
Engineering £126 m

Chemicals £942 m

Motor Vehicles £372 m

Other £393 m

Aerospace £818 m

1990
Total £5,299 m

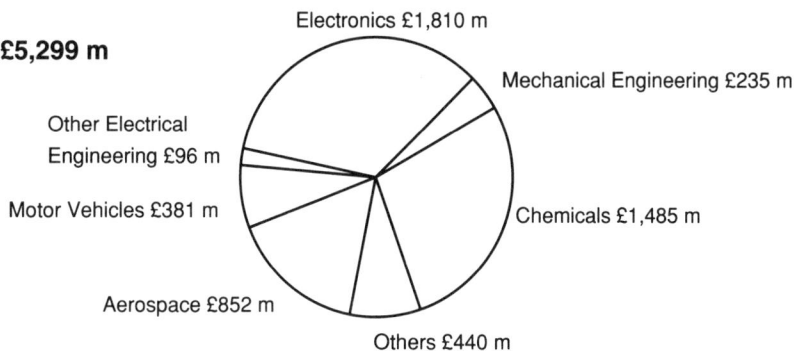

Electronics £1,810 m

Mechanical Engineering £235 m

Other Electrical
Engineering £96 m

Motor Vehicles £381 m

Chemicals £1,485 m

Aerospace £852 m

Others £440 m

1991
Total £4,729 m

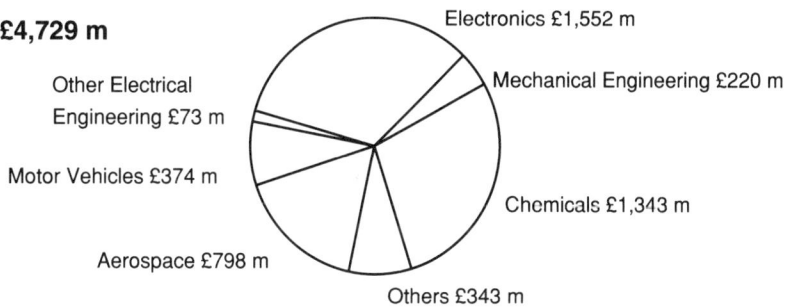

Electronics £1,552 m

Other Electrical
Engineering £73 m

Mechanical Engineering £220 m

Motor Vehicles £374 m

Chemicals £1,343 m

Aerospace £798 m

Others £343 m

Figure 38 R & D Performed in Manufacturing Industry (1985 Prices)

Source CSO

%

- - - Pharmaceuticals

—— All manufacturing

66 67 68 69 72 75 78 81 83 85 86 87 88 89 90 91

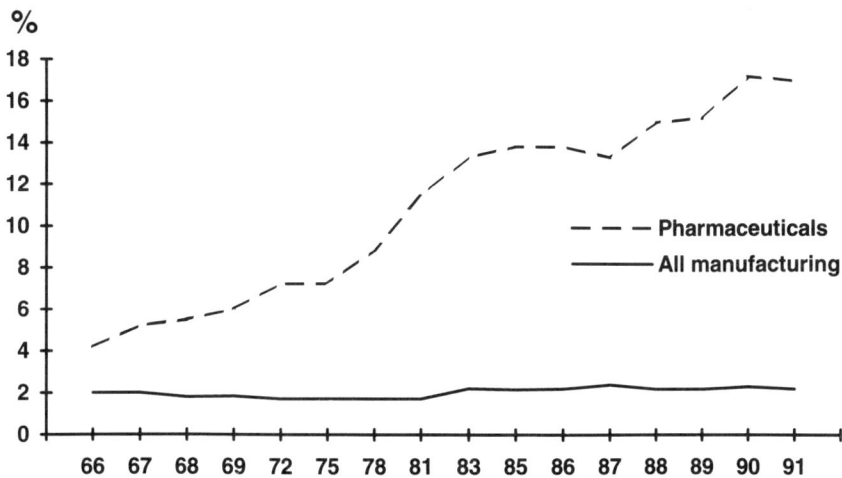

Figure 39 UK R & D Expenditure % of Gross Output

Source: CSO

59

Table 15 is taken from the annual "UK R & D Scoreboard" for listed companies published by Company Reporting Ltd [18]. It shows the latest financial years of stock-exchange listed companies.

Table 15 UK R & D 1992/93 - Listed Companies

	Current spend £bn	% annual change	per employee £	R & D % of sales	R & D % of dividend
All Industry (listed)	6,475	6	1,550	1.55	48.3
Aerospace	502	-8	2,540	3.19	392
Automotive	239	3	1,530	3.01	129
Chemicals	802	10	3,690	4.03	121.4
Electronical & Electronics	681	-5	2,230	3.85	75.6
General Manufacturing	493	23	850	1.06	33.5

In addition to these listed companies the R & D spend in the UK of the top 25 unlisted companies is a further £550m.

The 6% all industry percentage increase of the listed companies is encouraging and may indicate that the drop indicated in Figs. 35 and 38 for 1991 has in fact been halted. Particularly welcome is the 23% increase in General Manufacturing which tends to be low-tech activity. However, at only £850 R & D spend per employee there is still a long way to go in this sector.

The previous report[1] also showed the remarkable fact that in 1992 more than half the total UK industrial spend was accounted for by just 12 companies. In 1993 the top ten UK companies by R & D spend are given in Table 16.

This is 58% of the All Industry spend (including the unlisted companies). However these giants in the UK are not more than middling in global terms. ICI at the top of the UK list is actually only 47th in the world. The top spender is General Motors with £3,908m.

Table 16 1992/93 R & D spend - Top UK Companies

		Expenditure £m	Change on the year %	1990 position
1	ICI	647	9	1
2	Glaxo	595	25	4
3	Smith Kline Beecham	478	11	5
4	Unilever	461	8	3
5	Shell Transport & Trading	435	7	2
6	GEC	417	-4	6
7	BP	315	2	7
8	Wellcome	254	11	>12
9	BT	240	-1	9
10	Rolls-Royce	229	6	8
		4,071		

Last year UK companies spent more than twice as much on dividends as on R & D (Table 15, final column). In contrast, the top 200 international companies, ranked by R & D spend, reverse this ratio. This inevitably brings back concerns about "short-termism" in the UK. Are UK companies distributing too much by way of dividends and not retaining sufficient profit for re-investment in R & D? If so, how is this to be changed? See Conclusions, Section 9.

8.4 Innovation

8.4.1 Innovation - does it pay?

Innovation:
To make changes by successfully introducing something new

This dictionary definition is worth bearing in mind because it encompasses changes of all kinds and sizes both in technology and in management- all the way from a completely new product which sweeps the world like the compact disc to an incremental improvement in a traditional everyday component. The definition underlines the point that innovation is essential to economic progress.

According to an Innovation Trends Survey by CBI/Nat West[19] it pays to innovate. Nearly 75% of respondents reported profits from innovation investment made within the last three to five years. The sector breakdown is shown in Fig. 40. For manufacturing as a whole, 66% of all respondents perceived a profits improvement from innovation.

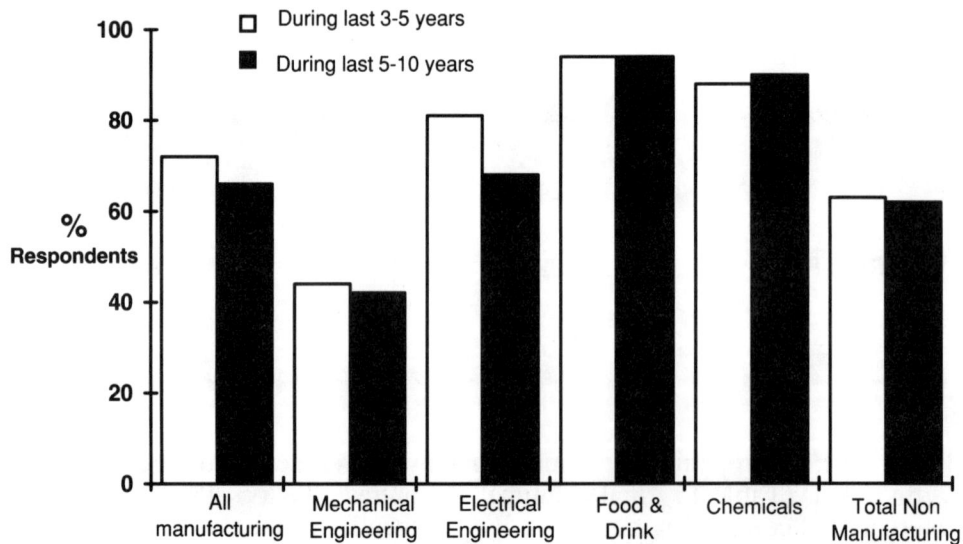

Figure 40 Profits Improvement from Innovation Investment

An indication of innovation in technology is revealed by the number of patents issued. Fig. 41 shows the comparison on the basis of foreign patents issued in the USA, the best indicator because all worthwhile inventions need to be protected there. Since 1980 the numbers in the UK and France have remained constant while those for Germany and particularly Japan, have increased.

62

Thousands

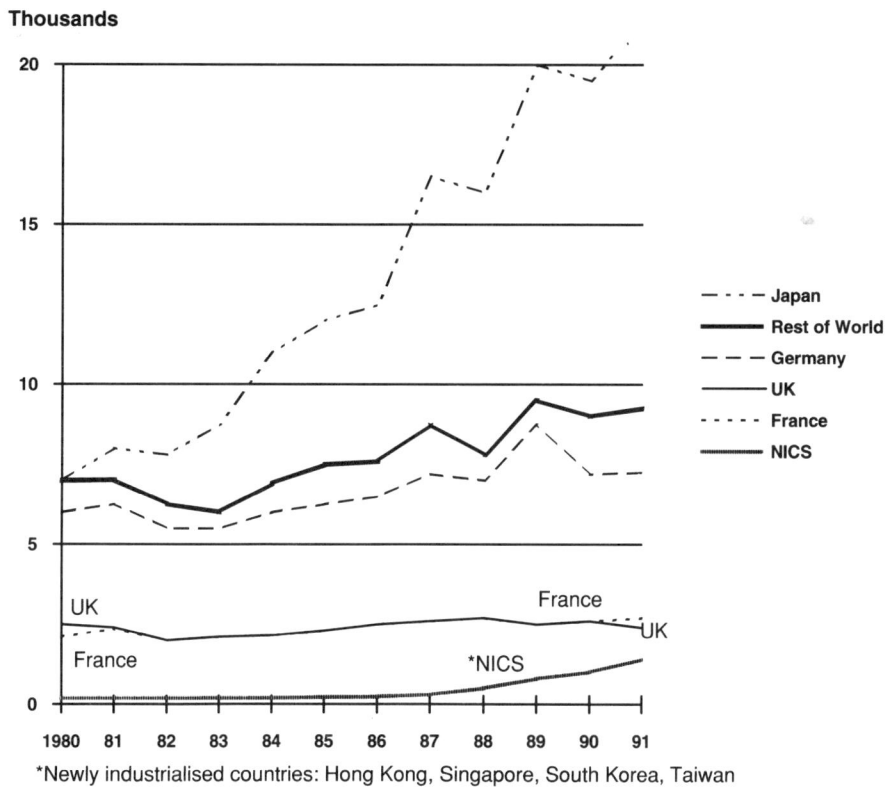

*Newly industrialised countries: Hong Kong, Singapore, South Korea, Taiwan

Figure 41 Foreign Patents Issued in the US

Source: DTI

8.4.2 A German - British Comparison

The CEST study[20] comparing Germany and Britain finds, in relation to innovation that:

- German firms use the domestic science and technology infrastructure intensively for the sourcing of technology.

- The contrast between the role of industry and the importance of manufacturing in the German and British economies is reflected at all levels. The need for British industry to improve its capability to innovate and its commitment to wealth creation is increasingly vital.

- Asked to name commercially successful innovations, both British and German industrialists put Personal Computers at the top of the list. However, after that there was an interesting divergence of view. The British responses were **radical innovations** such as mobile communication, videos, the Airbus wings, pharmaceuticals, optoelectronics, credit cards, body scanners, float glass and photocopiers. In contrast the German industrialists cited **incremental innovations in production technologies and environmental technologies.**

63

Maybe the main lesson here for UK companies is the importance of a continuous chain of relatively small improvements in design, production and maintenance made in-house. This is just as important as the occasional revolutionary new development, and perhaps more so.

The sources of innovative development are also different in the two countries (Fig. 42). The dominant source in Germany, more so than in Britain, is in-house R & D.

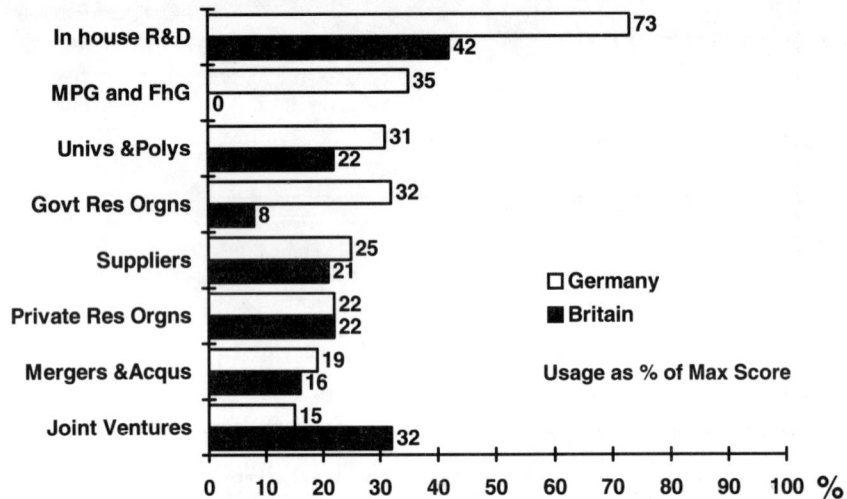

Figure 42 Current Sources of Technology: Britain and Germany

The dominance of in-house R & D in Germany is a reflection of the general attitude that German firms are reluctant to employ sources where the loss of control over technology is involved, as in joint ventures, or is a risk, arguably greater with foreign than domestic sources.

Another difference between the two countries is that there is no real equivalent in Britain for the general contribution made by the Max Planck (MPG) and Fraunhofer (FhG) Institutes.

What triggers innovation? Fig. 43 shows the external triggers by sector for Germany and Britain. The results are strikingly similar with customers at the top of each list. (The Environmental Concerns and Economic conditions in the British response were volunteered when participants were asked for "other triggers". In Germany no "other" factors were requested).

Germany

Britain

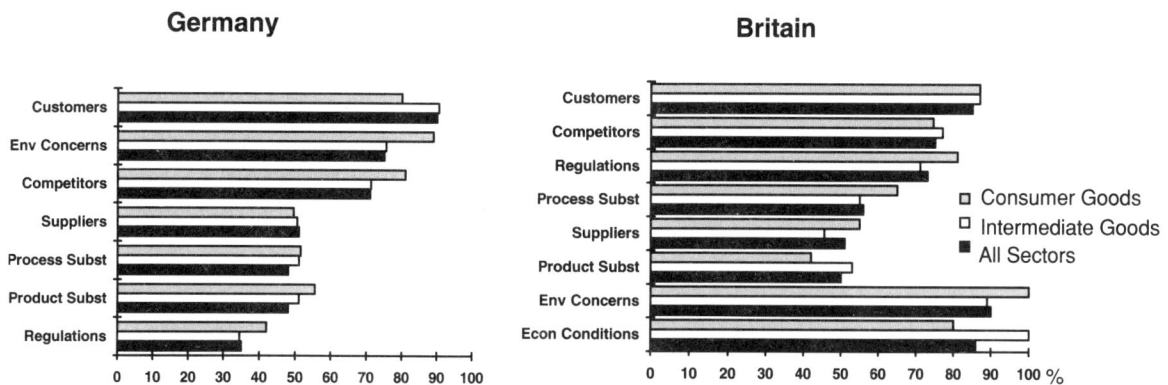

Figure 43 Innovation Triggers by Sector in Britain and Germany

8.4.3 How to Innovate Successfully

There is no magic formula for innovation. There are successful and unsuccessful examples in both the UK and Germany. The CEST report [20] summarises the conditions for successful innovation as:

- detailed appraisal of all relevant information

- evaluation of all the options

- maintaining the focus on delivering genuine benefits to the customer.

- undertaking sufficient R & D to ensure that performance and quality are not compromised

- quick utilisation of new technology

- maintenance of strong technological awareness in management

- capable creative staff at all levels in the organisation

- strong orientation to the customer

- commitment to product excellence and customer satisfaction

- product performance to meet the customer's requirements for use

- achievement of the correct balance of quality and price

- timely launch backed up by adequate capacity and delivery schedules.

A recent lecture on innovation by Harold Hayward[21] to the Manufacturing Division of the IEE presented Fig. 44 as a model for innovation.

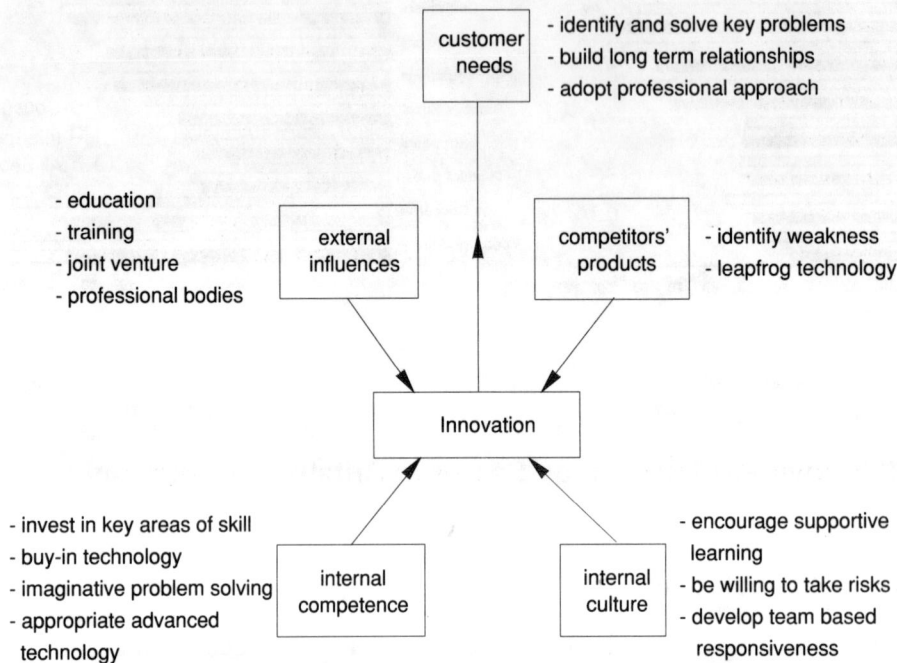

Figure 44 Model of innovation in a manufacturing company

He outlined the following as some of the requirements for an innovative climate in a small company:

- when entering a market with established competitors it is better to adopt an innovatory approach and leapfrog their technology than attempt to undercut them on price or delivery.

- the determination to succeed is more important in solving a technical problem than sophisticated resources and plentiful capital.

- to create a successful business, being a good engineer is not enough. One has to acquire competence in all aspects of the business, particularly commercial acumen.

- the in-depth analysis of a customers' problem can lead directly to a simple solution.

- quality in every aspect of company activity starts off with the desire and then the commitment to achieve it.

- investment in learning feeds business growth and innovation.

- it is possible for one determined individual to change and improve things locally, nationally and even internationally.

For the UK a number of important conclusions stem from these requirements for successful innovation. They are:

1. The underlying importance of education, training and retraining at all levels.

2. The significance of substantial R & D as the mainspring of innovation.

3. Emphasis should be placed on a national agenda for good management, such as that described in 8.5.4 as "Human Factors and Organisational Design".

The SERC has recently published a report [25] setting out the details of its Innovative Manufacturing Programme which is now being taken over by the new Engineering and Physical Sciences Research Council. One of the proposed initiatives stemming from the present report (Section 9.5) can be regarded as linking with the SERC programme and spreading the benefits to a much wider range of companies.

8.5 Management Style and Quality

8.5.1 UK Shortcomings

Shortcomings in UK management were explored in the previous UK-SOS publication[1] and reinforced in an unpublished DTI paper which is reported to say that although British industrial costs are competitive, the product base is weak and the country's main source of new products is imports. According to press reports it says the "hole in the heart" of British manufacturing is due to managers who, compared with their overseas counterparts, are poorly educated, ill trained and failing to turn technology into products that will win out in world markets. While managers have successfully eliminated chronic shop-floor over-manning they are often forced to take a short term approach which is hampering innovation.

8.5.2 Concurrent Engineering

A key aspect of technology management nowadays is that of "simultaneous" or "concurrent" engineering which aims to reduce the development times for new products from years to months. This has become known as "Rapid Prototyping and Tooling". Fig. 45 illustrates the way costs are committed and spent during product development.

The Figure also shows the benefits to be derived from concurrent engineering as assessed by Engineering magazine. The concept requires co-operation not just between design and manufacturing departments but also with marketing, suppliers and customers. Significant gains can often be achieved by "re-engineering" the business processes.

Product Development Cycle

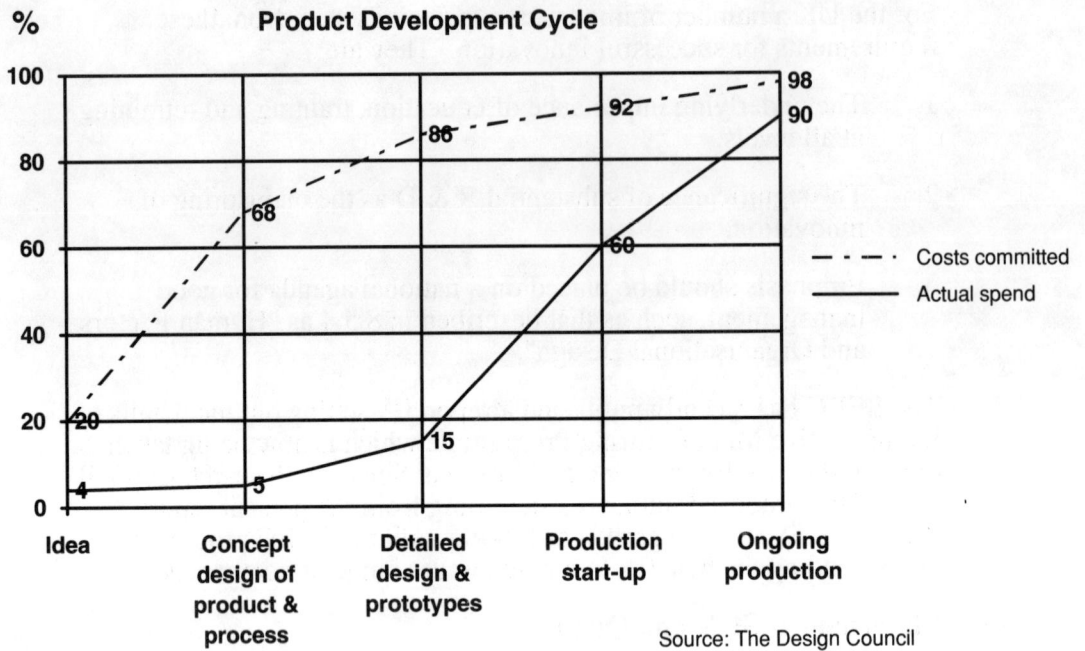

% (y-axis)

Costs committed (dashed line): 20, 68, 86, 92, 98
Actual spend (solid line): 4, 5, 15, 60, 90

X-axis: Idea | Concept design of product & process | Detailed design & prototypes | Production start-up | Ongoing production

Source: The Design Council

Benefits from Concurrent Engineering

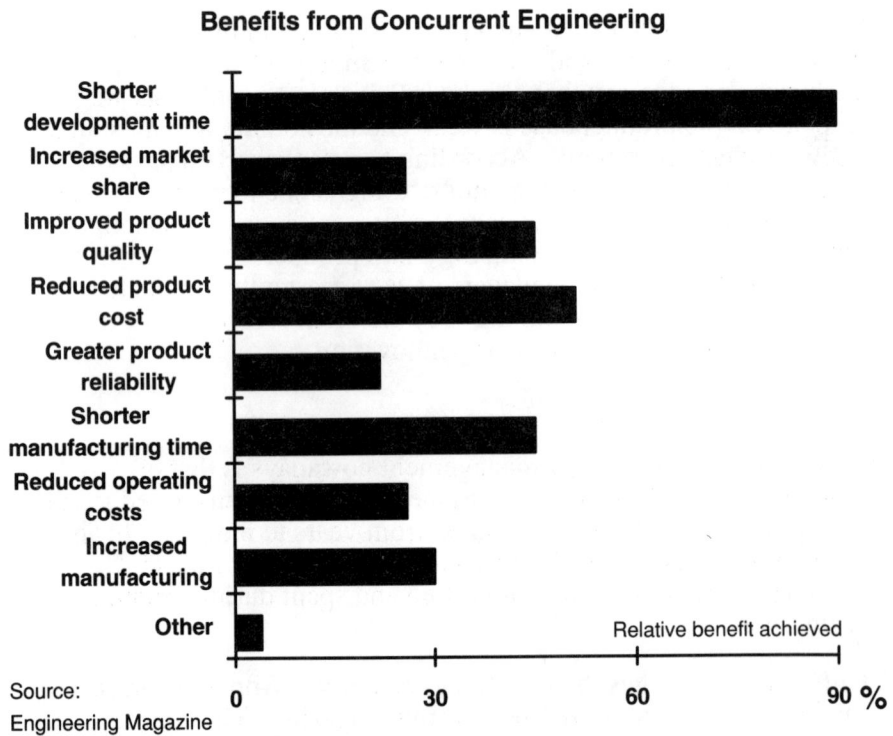

- Shorter development time
- Increased market share
- Improved product quality
- Reduced product cost
- Greater product reliability
- Shorter manufacturing time
- Reduced operating costs
- Increased manufacturing
- Other

Relative benefit achieved

0   30   60   90 %

Source: Engineering Magazine

Figure 45 Getting New Products to Market

8.5.3 Total Quality Management (TQM)

Quality is the name of the game so far as international competition is concerned. It is to the credit of the British Standards Institution that more than 20,000 firms are now registered to the BS5750 (ISO 9000) quality standard. Research carried out by Lloyds Register for Quality Assurance found that the majority of the registered firms benefited from their attainment and TQM, as complementary to BS5750, must be seen as a useful feature of global competitiveness.

Some smaller firms feel resentful of the need to register for BS5750 simply because, as they see it, they are compelled to go through a bureaucratic process as a requirement of the larger firms they supply. While respecting their difficulties, and recognising the case for a review of procedures, it is to be hoped that there is no retreat from the policy of promoting BS5750. Often it is the less good firms who strenuously object to any changes which they perceive only as an increase in costs while ignoring the potential advantages and opportunities which improved management control and increased efficiency bring. A similar negative attitude by these same firms is sometimes evident with regard to continuing education and training.

The Engineering Council and the Institutions could do more to encourage TQM, by giving it more prominence in their registration requirements and increased visibility in a new edition of SARTOR - Standards and Routes to Registration.

8.5.4 How to improve

What is to be done? An indication of current thinking comes in the interim report of the RSA's [23] "Tomorrow's Company" enquiry. It concludes that:

"If it is to achieve sustainable success in the demanding world market-place, tomorrow's company must be able to learn fast and change fast. To do this a winning company must inspire its people to new levels of skill, efficiency and creativity, supported by a sense of shared destiny with customers, suppliers and investors. In an inclusive approach of this kind success is not [wholly] defined in terms of a single bottom line, nor is purpose confined to a single stake-holder.

Tomorrow's company will understand and measure the value it derives from all its key relationships and thereby be able to make informed decisions when it has to balance conflicting claims of customers, suppliers, employees, investors and the community in which it operates. The company must be adaptive to its fingertips".

Conveniently also a 1993 report by ACOST [24] provides an agenda for improvement. It starts from the viewpoint that the future success of UK industry depends upon the effective commercial exploitation of science and technology. So it is essential that attention is paid to the interaction between people, the technology and the organisations in which they work. The report calls this interaction "Human Factors and Organisational Design" and proposes these strategic objectives:

- to recognise the reality that people, technology and organisations form strongly interacting systems.

- for senior management to recognise the importance of human factors and organisational design to the performance of their company.

- to make systematic application of existing knowledge an accepted procedure within organisations.

- to improve access to and dissemination of, researchers' and practitioners' knowledge.

- to implement a coherent national strategy for integrating education and training.

The ACOST report goes on to make a number of specific recommendations in pursuit of these objectives:

- companies should take action to integrate "human factors and organisational design" into the business process.

- Government departments and other public sector organisations should also consider how they may apply human factors and organisational design more effectively.

- the DTI should seek to stimulate wider awareness of good practice in human factors and organisational design.

- professional bodies for engineers, managers and accountants should build a requirement for knowledge of human factors and organisational design into their professional qualifications.

That the time may be right for such an initiative is indicated by another new publication from Ingersoll Engineers[26] on commercial planning and successful change. It reports some "encouragingly positive" findings in a survey of attitudes among managing directors and middle management teams. Unexpectedly it says (and in contrast to earlier reports):

(a) "The issue of two cultures no longer holds. There is an uncanny similarity of view between the managing directors and their middle management teams on all aspects of change. In fact middle management is shown not to be the point of resistance which [hitherto] so many believe it to be".

(b) "Professional issues of planning and communication emerged as the only remaining major barriers to continued improvement in the UK's industrial performance. It seems that simple misunderstanding is at the root of many difficulties".

Fig. 46 demonstrates the close agreement between Managing Directors and their managers on where changes will need to be accommodated and the relative importance of the key stages of change programmes.

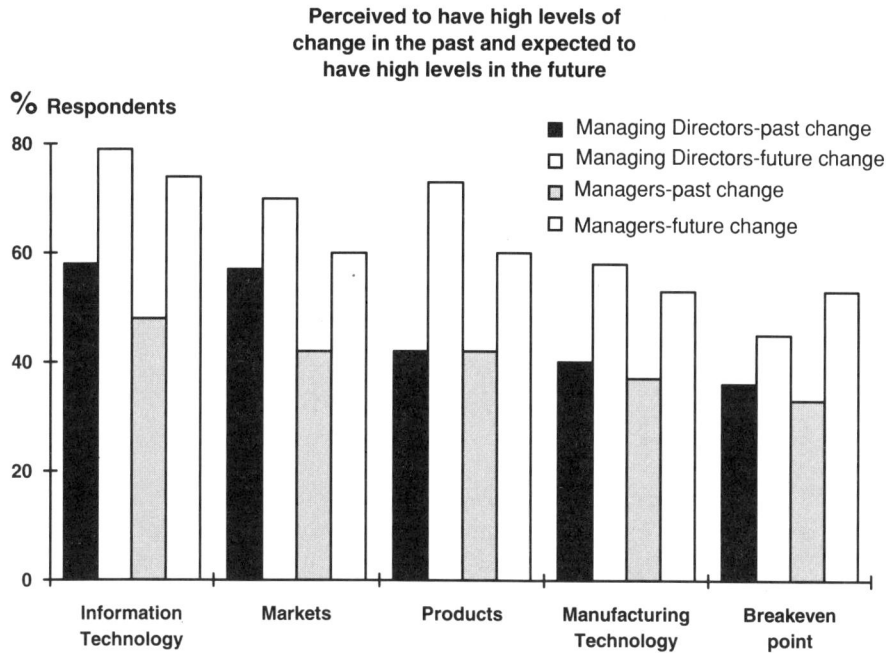

Perceived to have high levels of change in the past and expected to have high levels in the future

Legend:
- ■ Managing Directors-past change
- □ Managing Directors-future change
- ▨ Managers-past change
- □ Managers-future change

Categories: Information Technology, Markets, Products, Manufacturing Technology, Breakeven point

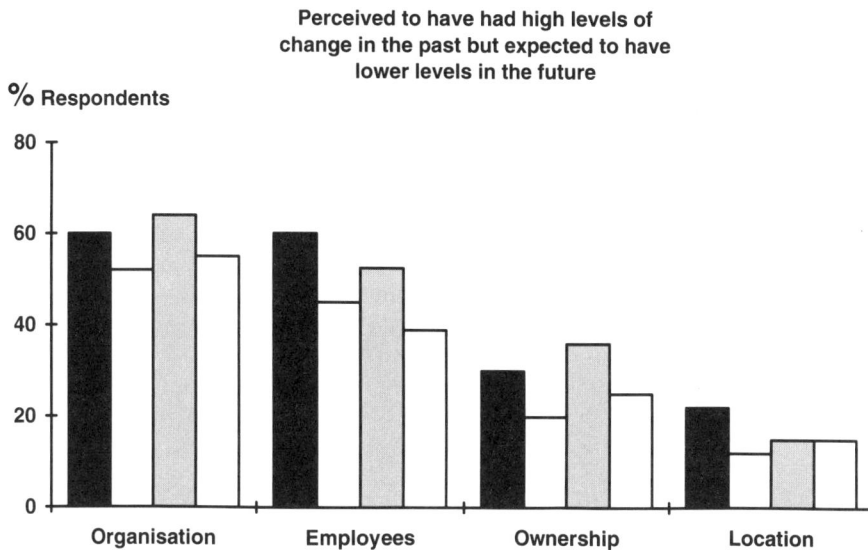

Perceived to have had high levels of change in the past but expected to have lower levels in the future

Categories: Organisation, Employees, Ownership, Location

Figure 46

Question: How would you describe the level of change in your company in the last 5 years and what level of change do you anticipate in the future?

8.5.5 Conclusion

If the DTI is right on the need, the RSA and ACOST are right on the formula for action and Ingersoll is right on a new consensus within management, then the elements may be in place for a step forward in the management sphere but a determined effort will be needed by the DTI and other agencies to stimulate the necessary understanding and action as described - See Conclusions, Section 9.

8.6   Size of company

8.6.1 Small and Medium-Sized Manufacturing Enterprises (SMMEs)

The question of company size, particularly in manufacturing has been explored in a recent companion publication "Small and Medium - Sized Manufacturing Enterprises - A recipe for success"[2]. Briefly the conclusions were:

- SMMEs (less than 500 employees) constitute 99% of our manufacturing units and employ about 45% of the manufacturing work force.

- They account for about 30% of manufacturing output so **on average** are less productive than large companies which provide 70% of output with 55% of the labour force. But, of course, some SMMEs are more productive than large companies.

- To help the development of SMMEs, particularly those medium-sized enterprises in the "growth corridor", lessons can be learnt from Japanese and German experiences, namely:

- In the case of Japan:

   the Japanese governments public stance, through its "Small and medium-sized enterprise basic law", which sets a framework of support and an annual review.

   improvements in technology in SMMEs

   improvements in productivity and added value

   benefits of inter-industry exchanges

   special financial arrangements for SMMEs via, for example, the Shoko-Chukin bank.

- And in the case of Germany:

   the tendency of manufacturing companies to be led by engineers.

   the system of training

   the provision for research and development in SMMEs via the AIF organisation which distributes funds for specific projects.

special financial arrangements for SMMEs through the KfW bank.

8.6.2 Measures to improve the performance of SMMEs

The problems highlighted in the previous reports[1][2] remain. We need to find means of encouraging prosperity and growth among firms in the 100-499 employee category, without damaging the activities of the larger firms.

SMMEs are a vital part of any manufacturing economy. They provide new ideas, act as suppliers to larger firms and some provide the seeds from which tomorrow's large enterprises will grow.

Three main barriers to growth of SMMEs can be identified - technology, skills and finance.

1. In technology, SMMEs need to maintain a competitive edge for survival. One immediate need is for their owners and directors to be given better access to projects being undertaken in Government laboratories so that the fruits of tax-payers' money can be used as widely and rapidly as possible to benefit the UK. Also, further consideration needs to be given to encouraging R & D in SMMEs. The German AIS model commends itself for further study.

2. In skills, (not just in technology but also marketing and management know-how) international comparisons repeatedly show that, apart from the top technical level, the UK work force is seriously under qualified. This is a particular problem for smaller companies because the large companies have first pick of the high-calibre talent from an inadequate pool, leaving too few for the SMMEs. Consequently improvements in education and training must remain at the top of the national agenda.

3. In finance, the most frequently expressed of the limiting factors, more imaginative approaches are needed. Measures announced in the November 1993 budget are helpful in some respects to small companies, but the twin successors to the Business Expansion Scheme do not meet the basic need for the owner or entrepreneur to be able to borrow more cheaply without losing control of the company. We need a specific Manufacturing Expansion Scheme and a new arrangement to enable SMMEs in the "growth corridor" to be able to borrow money at similar rates to those available to large organisations - see 8.7 below and Conclusions, Section 9.

These three basic actions must be stimulated by Government, but Government measures alone will not enable SMMEs to flourish. Each company, if it wishes to expand, needs to be aware of DTI and EU initiatives. There are some excellent DTI schemes to assist SMMEs such as the development of the Business Link Network (One Stop Shops) with the TECs and Chambers of Commerce, but their penetration is too low. The SMME report[2] gives 25 measures for each company to take. A useful move in the last few months has been made by the CBI in launching the Competitiveness Forum under the auspices of its National Manufacturing Council. The Forum is a network of ongoing and varied consultancy events designed to put the expertise of Britain's world class businesses at the disposal of a broad spectrum of manufacturing companies. It intends to "offer a series of opportunities to companies across manufacturing sectors to learn from the best source possible each others practice".

On the Forum's agenda will be production processes and engineering, new product development and technology, bench-marking and performance measurement, together with broader issues of business strategy, management and marketing. Involvement and support is coming from the DTI.

Table 17 Lifetimes and stock of premises and equipment

	Service life of industrial premises (years)	Equipment life (years)	Capital stock/worker*
Britain	60	24	100
USA	31	17	173
France	37	17	148
W Germany	41	15	194
Japan	43	10	131

* UK = 100
Source NIESR

8.7 Finance, Investment and Government Policy

In the previous UK-SOS report[1] the inter-relationship of investment in industry, education and the skills base was examined. It was pointed out that the capital spending per employee in UK manufacturing was only about 60% of our main EU competitors and 40% of Japan's. This relative under-investment is re-emphasised in Table 17 and Fig. 29 which show no sign of improvement.

74

Low capital investment is not the only negative factor. As already shown in Table 15, UK expenditure on R & D is about half that of dividends to shareholders while in the 200 top companies in the world, that ratio is reversed.

Why is this? Why do we appear to be consistently near the bottom of these "league tables" with the result that we have difficulty in closing the trade gap in manufactured goods and in being sufficiently successful in import substitution?

It was pointed out in UK-SOS [1] that the pattern of share holding in UK companies is markedly different from that in W Germany, USA and Japan. The main difference is the percentage of shares held by pension funds and insurance companies, being 49% in the UK, 25% in the USA, 16% in W Germany and only 7% in Japan.

Pension funds of all types require a high return and high growth. In the UK pension funds can reclaim tax on the net dividends they receive and this additional bonus makes it very difficult for high-tech manufacturing companies to reduce dividends in favour of R & D or of capital investment when the pension funds can easily switch their money to other investments which are less R & D and capital intensive and so can maintain dividends more easily.

The present system puts manufacturing enterprises, which are invariably R & D and capital intensive, at a clear disadvantage so it is extremely important that this particular playing field is levelled by the Government granting better tax concessions for innovation, including R & D and for capital spending. To repeat the previous report UK-SOS[1] "Investment in technology is the means by which an economy builds its productive base and its output potential".

To foster the expansion of our manufacturing base, more imaginative approaches are needed for example, a manufacturing expansion scheme, which gives tax breaks specifically for investors in manufacturing enterprises. Also, picking up the point made in 8.6.2, we need a special bank to support SMMEs.

This function can no longer be left only to the traditional banking system because they have many larger interests to serve. Experience shows that SMMEs can never enjoy high priority with them. Neither can venture capital organisations or "business angels" fill the gap because they expect too high a rate of return and may require part ownership of the business which many proprietors of SMMEs are unwilling to concede.

So, a new bank is needed which would have SMMEs as its top and sole priority. Its function would be to lend to approved customers on the same terms as available to larger companies so levelling the playing field in this respect and giving SMMEs a better chance for rapid growth. But money is not all. This service, as in Japan, should include specialist information and advice from trained bank staff employed for that purpose without the strings which entrepreneurs dislike, of equity sharing or the presence of non-executive directors. The experience in Germany and Japan is that the provision of expert advice as a service both before and after a loan is granted reduces risks all round and that such a bank once established would not be a drain on public funds.

PART 3

CONCLUSIONS

9. Challenges and Responses

9.1 The Broad Picture

The past two years have seen profound international changes which directly affect the prospects for UK manufacturing and to which appropriate responses must be found. The pace of change is staggering. In Europe the break up of the USSR and the re-unification of Germany; across the Atlantic, the launch of the North American Free-Trade Area (NAFTA) and the Latin American equivalent, MERCOSUL; the establishment of the Asian Pacific Economic Co-operation (APEC) organisation; the possible ending of the Japanese "miracle"; the rise of the Chinese economy and the "little dragons" of Singapore, South Korea, Taiwan and Hong Kong along with other Pacific Rim countries; the success of the Uruguay round of the GATT negotiations. All these set the agenda for years to come and add up to significantly greater challenges for UK manufacturers which must be met if we are to survive as a trading nation.

The nature of the world today is that no single country has the freedom of manoeuvre to solve its economic problems independently. Dominant influences in the developed countries are the cheap labour in developing countries and the ability to transfer huge sums of money between continents at the touch of a few buttons.

Consequently for the UK and other European nations, there is no alternative (short of an unthinkable retreat into an isolationist EU), to take whatever action is needed to keep abreast of developments. British companies must scour the world for contracts, giving customers the products and services they want. Here is a summary of the international and domestic challenges facing the UK, followed by suggested responses and a set of four linked national initiatives as part of a Manufacturing Expansion Scheme.

9.2 International Challenges

9.2.1 The EU

- the generally lower productivity of the EU generally and the UK particularly compared with the USA and Japan

- the EU trade deficit with the rest of the world

- the strength of the Franco-German axis which tends to dominate the EU

- EU over capacity in major industries such as steel and chemicals and agriculture

- the enlargement of the EU with potential benefits, but new problems too.

9.2.2 Eastern Europe

- uncertainties about the future stability of the system

- the huge backlog of industrial, environmental, consumer product and infrastructure needs.

9.2.3 The USA

- tremendous industrial muscle in spite of recent setbacks

- plans to invest heavily in research and development

- de-regulation of hitherto restricted export products, so increasing USA trade advantages

- North American Free Trade Area (NAFTA) may reduce costs and increase jobs in the USA

- determination to liberalise trade with Japan and penetrate the Japanese market.

- ambitions to play a leading role in the Pacific Rim

- sees massive advantages for its industries and agriculture through GATT.

9.2.4 Japan

- in spite of current downturn and possible ending of the economic "miracle", great strengths in manufacturing, R & D and methods of finance

- taking leadership in bi-lateral arrangements with other Asian countries

- cushioned by huge trade surpluses with the USA and EU giving scope for measured response to present economic difficulties

- developing links with China which, in spite of ambiguities in relationship, could present a formidable combination

- well established methods for allowing small and medium-sized manufacturing companies to borrow long term at low rates.

9.2.5 The Pacific Rim

- a medley of countries with different policies, attitudes, rates of development and aspirations. Some, like South Korea, are very successful industrially and many have low labour costs

- the new APEC organisation which for the first time involves Pacific Rim Governments in a trade group

- a challenge to the UK and EU to compete with the USA and Japan in trade with the diverse countries of SE Asia

- China as the emerging great power with a high rate of growth. The UK has a growing trade deficit with China.

9.3 Domestic Challenges

9.3.1 The 4 Key Indicators

The UK does enjoy certain advantages by reason of its economic policies, but in terms of the 4 key indicators:

Output - the increase between 1975 and 1991 compares unfavourably with the USA, Japan and Germany.

Productivity - despite improving at a greater rate than competitors, productivity still lags in absolute terms.

Share of World Trade - has continued to decline and may now be below our "fair share" figure in proportion to population.

Import Penetration - the trade balance worsened in 1992 and 1993 particularly with non-EU countries. The GATT round could accelerate the move of manufacturing to low labour cost countries.

9.3.2 The 6 inputs

The 6 inputs are so called because they are under UK domestic control:

Education and Training - by international standards, our population is still under-educated and trained particularly at the lower levels. Considerable problems remain on NVQs, training and the funding of higher education. Engineering degree courses need to be reviewed.

Research and Development - there was an actual reduction in 1992 over 1991. Our best firms have a 2:1 ratio of distributed profits to R & D whereas in the 200 best internationally rated firms this balance is reversed. The White Paper "Realising our Potential" is helpful but will need energetic and determined follow-up.

Innovation - recent studies have assisted the identification of successful routes to innovation but have not yet been widely applied. More emphasis is needed on incremental innovation to complement new product innovation, carried out in-house by firms of every size and type.

Management Style and Quality - still leaves much to be desired on average. Known lessons are not yet applied widely enough. Concurrent engineering and TQM insufficiently exploited.

Size of Companies - most of the 50 points put forward for improvement of SMMEs in the previous paper[2] remain to be assimilated into the culture. Financing problems need to be solved for those SMMEs in the growth corridor.

Finance, Investment and Government Policy - our plant is relatively old and we still lag behind in capital investment. The taxation system gives inadequate recognition to vital high technology industries which depend upon heavy investment in R & D and equipment.

9.4 The UK Response

9.4.1 General Responses

- we must not be deflected from education and industrial policies which encourage the continued improvement in productivity and reduction of costs. Without this we will continue to suffer economically. Although our recent record shows *percentage* improvements in productivity, greater than other developed countries, our absolute level is still below the USA, Japan and Germany. To catch up we need to exceed their rates of growth by at least 1% per annum for each of the next 10 years.

- although it is true that we export a greater amount per head than either Japan or the USA, we lag behind other EU countries on that measure. In the main fields of engineering our manufacturing exports still have a positive trade balance but they have to pay for a huge volume of imports in fields such as food, textiles and paper, where we run considerable trade deficits. So as well as stimulating exports, we need to pursue ways of reducing the need for imports by manufacturing more competitively to satisfy our own needs.

- continue to make the UK an attractive location for foreign investment and companies, especially where R & D facilities are included in the development.

9.4.2 Area-Specific Responses

In the EU

- increase pressure on other EU countries to reduce their high level of subsidy to manufacturing industry - particularly Italy, Germany and France. Also on deregulation such as that in the air transport industry. Expand knowledge and know-how among UK companies to gain public sector contracts in other EU countries.

- support the further development and co-ordination of EU research programmes with emphasis on near market efforts and generic technologies.

- Eastern Europe - it is in the interests of the UK that stability prevails. Helped by careful investment in manufacturing and infrastructure. By western standards investment in Eastern Europe is cheap and could bring longer term benefits to the UK. Explore collaborative agreements to use UK expertise in combination with host country resources.

Non-EU countries

- the fact that most of our trade is now with Europe is, in one sense, encouraging. But it is, in part, an arithmetical consequence of the persistent deficit with non-EU countries. So more attention needs to be paid to increasing the exports to and reducing the imports from, countries outside the EU. The success of the GATT round must be exploited to the full in increasing exports outside the EU.

The USA

- the redoubled USA effort in civilian orientated R & D requires a similar response from ourselves and other EU countries if we are not to be left in a technological backwater. The efforts of the USA through NAFTA and other channels will lead to reduced costs and more competition for Europe, so efforts to reduce our own costs and improve efficiency cannot be relaxed.

Japan

- be conscious that Japan is also facing economic and industrial difficulties which paradoxically could lead to even sharper competition.

- continue to learn from Japanese manufacturing methods and their R & D techniques in bringing new products to the market. Join USA efforts to open up the Japanese home market to imports. Learn from Japanese ways of providing growth finance and adapt them to UK culture.

The Pacific Rim

- do not allow the USA and Japan to have it all their own way in the Pacific Rim. Perhaps Commonwealth links could be re-invigorated and used more effectively for this purpose. We are perceived as less of a threat by the Pacific Rim countries than either Japan or the USA and this could work in our favour. China is a special case and we must take clear and effective steps to ensure that our manufacturers are not excluded from the enormous potential of their growing economy.

9.5 A MANUFACTURING EXPANSION SCHEME:
Four National Initiatives And Their Underpinning

Much remains to be done to create a long-term growth environment. Despite some catching up in productivity and some favourable economic indicators, there is little sign of a reduction of our balance of payments deficit and it has been shown time-and-time again that the only way to meet the overseas challenge is to increase our exports of manufactured goods. Otherwise we will not be able to pay for necessary imports of food and raw materials, let alone the minimum of manufactured goods we need from other countries.

The necessary responses to the challenges set out above can be articulated through just four linked National Initiatives in Management, Technology, Finance for industry and Innovation, all underpinned by improved Education and Training. Through them we can promote high-tech growth so that our output, productivity, share of world trade and balance of payments all improve.

These four initiatives as part of a Manufacturing Expansion Scheme, will not produce a "miracle cure". There is no such thing. But taken together they could set a steady path to expansion of our manufacturing base.

a) **A MANAGEMENT INITIATIVE**
People and Organisational Design

The DTI has identified lack of management skills as a major deficiency of UK industry. Recent RSA [23] and ACOST reports [24] have studied the interaction between people, the technology and the organisations in which they work and have identified a number of strategic objectives which need to be addressed if UK science and technology are to be successfully exploited commercially. Among the objectives are:

- widespread recognition of the reality that people, technology and organisations form strongly interacting systems.

- recognition by senior management of the importance of human factors in organisational design to the performance of the organisation.

- systematic application of existing knowledge in the field making it an accepted part of procedures.

Another recent study [26] has revealed that a new consensus appears to exist between senior and middle management. If the DTI is right on the need, RSA and ACOST on the strategic objectives and there is indeed a new consensus, then the time is right to launch a new national initiative for the improvement of management and employment generally, under the title "People and Organisational Design".

To achieve the objectives, actions are needed by companies, Government departments and professional bodies:

- companies should take action to integrate "people and organisational design" into the business process by ensuring that these are reflected in career development patterns and training programmes which encourage the acquisition of a broad range of skills.

- Government departments, and other public sector organisations in a similar way should seek to apply the principles more effectively and also seek to promote the further penetration of the DTI's "Manufacturing Organisation, People and Systems" (MOPS) scheme which could substantially raise awareness. The DTI should review all aspects of its "Managing into the 90s" programme to stimulate wider dissemination of good practice in advanced manufacturing technology.

- professional bodies for engineers, managers and accountants should build into their professional qualifications a requirement for knowledge of people and organisational design.

Such a programme would also include the lessons in the CEST [20], and Hayward [21] studies on how to innovate successfully.

This whole initiative is not so much a matter of additional resources as of mobilising existing activity and ideas behind an achievable national aim. A high level body, new or existing, should be given the responsibility of spearheading the co-ordination of education and training efforts in this area and advising on the distribution of available funds.

b) **A TECHNOLOGY INITIATIVE**
The Agile Factory

The polarisation into major trading blocs, comprising the EEA (including the EU), NAFTA and APEC, with freedom of trade within and between them via successive rounds of GATT, means a continuing shift of manufacturing activities to low-wage countries and so there will be a continuing pressure on our manufacturing base, particularly in the low and medium technology sectors.

What is the answer? Part of the reply should consist of the development of what is called the "agile" factory, rapidly responsive to changing circumstances and to customers needs. This implies the capacity for rapid prototyping and tooling and, where relevant, a continually increasing level of flexible automation, not just of component manufacturing which is often highly automated already, but of assembly processes too. In combination with a highly-trained multi-skilled workforce, this would counteract part of the advantages of overseas cheap labour. Improvement will be achieved through the wider application of concurrent engineering, total quality management and, where necessary, "re-engineering" of the business organisation itself.

The necessary R & D will, in itself, be a valuable export product and will assist the lifting of the whole manufacturing effort. Closer co-operation on the lines of an "extended-family" will be needed between factories which are members of the same group and, between them and their sub-contractors.

Some of our best manufacturing facilities already include elements of the "agile factory" concept. Examples are to be found in our best electronics companies and in the ability of Lotus to bring a new racing car from concept to race-track in 6 months by a process heavily dependent on CAD.

But a much deeper penetration into our manufacturing culture is required. The new Competitiveness Forum, launched by the CBI, enabling firms to share good practice and the learned society activities of the engineering institutions could have an important role in this effort . The agile factory as a widely promoted initiative is also a logical follow-up to SERC and DTI efforts to promote world-class manufacturing.

Good examples of industries where there is scope for the UK to expand its manufacturing base with the agile factory concept are to be found in such diverse areas as machine tool manufacturing, mechanical transmission systems, finished textiles and generally in electronic manufacturing and sub-assembly. All these have to cater for long and short runs to satisfy customer needs which differ in markets across the world and which change rapidly with time.

This project would require close links to be established between design and automated assembly functions and would also link with the Technology Foresight and other proposals in the White Paper "Realising our Potential".

c) **A FINANCE FOR INDUSTRY INITIATIVE**
A new arrangement for Small and Medium-Sized
Manufacturing Enterprises (SMMEs)

We have no equivalent to the KfW in Germany (established 1945) or the Shoko-Chukin bank in Japan (established 1936) which enable SMEs to borrow at much the same rates as large organisations, so levelling the playing field in this respect and enabling vigorous medium-sized companies with good track records to grow more rapidly into larger ones.

The matter can no longer be left to the traditional banking system based upon the large banks. Experience shows that, whatever their goodwill, the small and medium sized sector can never rank high in the banks' priorities. They have much more important activities occupying their attention and, even with good intentions, the expertise of their staff at street level will always be lacking. Venture capital or "business angels" cannot fill the gap because, even with the recent budget changes, the returns they require are generally too high and they are not always in a position to offer the necessary long term support and advice.

New provision is clearly needed, particularly directed at those companies in the "growth corridor" employing 10 to 50 people. What the owners of SMMEs really want is to be able to borrow money more cheaply, on a par with the rates available to large firms. This is true even with the recent lower rates. But money is not the only requirement. Entrepreneurs also need to have on-tap a reservoir of expert information and advice without the "strings" of equity sharing or acceptance of non-executive directors. One solution would be to create a special bank on the lines of Japan's Shoko-Chukin, which not only advances cheap fixed rate loans to properly vetted applicants but also provides the necessary intellectual support, constantly available with expert staff. Risk assessment and management should be a continuous part of the process both before a loan is granted and afterwards. In this way the risks are minimised all round.

The basic financing of such a bank would have to be arranged. Assuming that the traditional banks would not co-operate in supporting a bank system which is so radically different, the Government would have to act as final guarantor. However if the experiences of the Japanese bank and of the KfW in Germany are typical then, after initial pump priming, any necessary Government guarantees would not have to be called upon.

d) **AN INNOVATION INITIATIVE**
Taxation of Innovative and Capital Intensive Enterprises

Innovation includes entirely new products and also incremental innovation of existing lines. Its importance in manufacturing industry has been demonstrated on many occasions - see Section 8.4 of this report. However, the present system of taxation puts our high-tech industries at a serious innovative disadvantage.

Let us compare two companies with similar turnover. One is high-tech and characteristically needs to invest heavily in innovation (including R & D) and in capital equipment. The other has fewer of these needs in order to compete.

The first type of company has to carry salaries and overheads of highly-paid scientific and technical staff. Even though these count as a business expense they reduce profit like any such expense. Also, only about 69% of investment in capital equipment can be written off over four years with our present system of allowances.

Yet the high-tech company has to pay a similar level of dividend as the other type of company otherwise it cannot attract shareholders, particularly pension funds, which are such a uniquely important feature of UK stock holding compared with other major countries. Pension funds can even reclaim tax on the dividends they receive, so that maintaining the level of dividends is doubly important to them.

To remedy these disadvantages suffered by high-tech companies, a new scheme is needed which would specifically acknowledge (i) the underlying importance of innovation and (ii) the need for world competitive manufacturing companies (of whatever size) to continuously update their capital equipment.

(i)   Additional taxation allowances for particular functions are admittedly a notoriously difficult area because of the problems of definition and ensuring that companies do not claim for non-qualifying activities. But perhaps in the case of innovation the difficulties are exaggerated if we focus on how it is actually achieved.

Innovation includes R & D, design, manufacturing and marketing. It therefore depends crucially upon the employment of qualified people. A very recent NIESR report [27] closely links the employment of highly-qualified staff to success in product and process innovation in Germany as compared to the UK. It would be logical to base an additional tax allowance on the total cost of staff possessing appropriate professional qualifications in engineering, science or marketing who are employed in each manufacturing company. As a starting point, it is suggested that 150% of the costs of such staff plus an agreed percentage for overheads and support should be allowed for tax purposes. This would be an easily workable system because such staff are readily identifiable by their qualifications. There would be a number of immediate beneficial effects; innovation would be encouraged and the scheme would motivate staff and employers alike. In line with Government policy, the importance of proper qualifications could be stressed including postgraduate and degree level, but higher technician qualifications too, including, for example, NVQ level 4 as part of the team effort. Also the initiative would help to combat the anti-industrial culture in this country by tangible and public recognition of the importance of industrial innovation and of the staff engaged on it.

It is not intended that the additional tax allowance should be used to pay higher salaries - market forces will take care of those. Rather it should enable the employment of additional staff and resources to enhance innovation.

This initiative would link with and deepen the penetration of the SERC Innovative Manufacturing Programme [26] enabling a broad spectrum of companies, small and medium-sized as well as large, to benefit by providing the necessary ongoing resources.

(ii)   The arguments for 100% capital allowances in year of purchase, or at a rate chosen by companies themselves, are persuasive. They would assist our companies to re-equip at the same rate as our competitors and raise to their level the amount of capital employed per worker.

The cost of these changes to the Treasury is put at £2-3bn annually but should be easily recovered through the taxation of increased profits on a broader manufacturing base.

This initiative would help UK manufacturing to establish a trade record of being a reliable, steady earner. That way shareholders, including pension funds, would be encouraged to invest.

e) **THE ESSENTIAL UNDERPINNING**
   Improve and Integrate Education and Training

Better education and training remain fundamental and continuing improvement in these areas is necessary to underpin the four initiatives in manufacturing. Shortages of well-qualified people hit all enterprises at any and every level. But the small and medium-sized ones suffer most because the large enterprises have first choice from the pool. We need to simplify and redesign the provision to create a properly integrated and readily understood system of education and training by:

- maintaining the impetus on the National Curriculum and putting better teaching of mathematics and science as a main priority. In the Technology syllabus, implementing the latest (September 1993) proposals on simplification and developing towards a clear central theme.

- ensuring that NVQs are taught and examined in such a way that, on top of demonstrating immediate competence in a particular activity, they lead to a work force with flexible and competitive skills. This means more attention to the underlying knowledge base.

- explaining more clearly to employers and the public the plethora of qualifications (GCSE, NVQ, GNVQ, A-Levels, A/S-Levels, RSA, C&G, BTEC etc.) and routes which are extremely confusing to all but the initiated.

- ensuring TECs can fulfil the "Enterprise" part of their mission in co-operation with Chambers of Commerce and other interested organisations.

- making firm decisions on the future expansion and funding of Higher Education so that universities can plan ahead to meet the need.

- expanding tax relief for those in job-related continuing education and training. Professional institutions should promote recognised continuing professional development of their members.

- launching a programme to ensure that all young people acquire proficiency in a second language by 2010. This would remedy a notorious deficiency in our overseas dealings .

- launching a national debate on a rationalisation of the whole education and training system.

10. REFERENCES

1. "UK Manufacturing - A Survey of Surveys and a Compendium of Remedies", The Institution of Electrical Engineers, May 1992 (referred to as UK-SOS).

2. "Small and Medium-Sized Manufacturing Enterprises - A Recipe for Success", The Institution of Electrical Engineers, July 1993.

3. "Industrial Policy in OECD Countries - Annual Review 1992", OECD, December 1992.

4. "Is Manufacturing still special in the New World Order"? Richard Brown and Deanne Julius. Amex Bank Review 1993, Oxford University Press. pp 6-20.

5. OECD Economic Surveys 1991/1992, Japan

6. "Making it in Britain", CBI, Autumn 1992.

7. Treasury Bulletin, Vol. 4, Issue 2, Summer 1993.

8. National Institute Economic Review, No 142 November 1992.

9. "Made in Britain - The true state of Britain's manufacturing industry" - IBM Consulting Group/London Business School, Summer 1993.

10. "Educational provision, educational attainment and the needs of industry: A review of research for Germany, France, Japan the USA and Britain" Andy Green and Hilary Steedman, National Institute of Economic and Social Research, Report Series No. 5, 1993.

11. "Learning to Succeed", Report of the National Commission on Education, Heinemann, November 1993.

12. "Local Empowerment and Business Services: Britain's Experiment with TECs", R J Bennett, P Wicks and A McCoshan, UCL Press, January 1994.

13. "Review of Engineering Formation - A discussion document", The Engineering Council, August 1993.

14. "Engineering Requirements in 2010: A scenario for skills", M White, Policy Studies Institute, June 1993.

15. "Engineering Recovery", Howard Davies, 4th Bridge Lecture, City University, February 1994.

16. "Realising our Potential - A Strategy for Science, Engineering and Technology", HMSO, May 1993, Cm 2250.

17. Annual Review of Government Funded Research and Development 1993, HMSO

18. "Research and Development Scoreboard 1993" David Tonkin of Company Reporting Ltd . Available from DTI. (Also published in The Independent 9th June 1993 pp 28-29).

19. CBI/NatWest Innovation Trends Survey, NatWest Technology Unit, Issue 2 March 1992. Also CBI Manufacturing Bulletin, No 4, May 1993.

20. "Attitudes to Innovation in Germany and Britain: A Comparison". Centre for Exploitation of Science and Technology (CEST), July 1991.

21. "Innovation - the key to success in manufacturing", Dr Harold Hayward, IEE Manufacturing Division, October 1993.

22. "How to make Re-engineering Really Work", G Hall, J Rosenthal and J Wade, Harvard Business Review, November-December 1993.

23. "Tomorrow's Company: The Role of Business in a Changing World" Royal Society of Arts, Interim Report, February 1994.

24. "People, Technology and Organisations - The application of Human Factors and Organisational Design". Advisory Council on Science and Technology (ACOST), HMSO September 1993.

25. "Creating Confidence - Communication, Planning and Successful Change", Ingersoll Engineers, September 1993.

26. SERC report on the Innovative Manufacturing Programme, March 1994; Science and Engineering Research Council, Swindon

27. "High-level skills and industrial competitiveness - postgraduate engineers and scientists in Britain and Germany". G Mason and K Wagner National Institute for Economic and Social Research, Report Series No 6 February 1994.

General: Surveys and reports in Financial Times, The Times, The Economist, The Design Council and Engineering Magazine.

11. BIBLIOGRAPHY

Reports on the UK economy have continued to be published apace in 1992 and 1993.

Among the more important which have influenced the present paper but have not been directly quoted are:.

(a) **"Making it in Britain"** CBI, Autumn 1992 : reworks the arguments and includes a new checklist for action similar in layout to the "Compendium of Remedies" in the UK-SOS [1] report. Its recently formed National Manufacturing Committee has set the following targets:

- productivity increases of at least 5% a year through the rest of the decade, a higher rate than that achieved in the 1980s

- a doubling of investment per employee in plant and machinery and further real increases in investment in skills, innovation and marketing

- the achievement of an extra 1% of world trade worth £10bn a year to exports and a drive towards import substitution.

(b) **"New Product Development"** Design Council, October 1992 : describes case histories of a number of successful new products from (mainly) small or medium-sized manufacturing companies.

(c) **"Can De-industrialisation Seriously Damage Your Wealth?"** Professor N F R Crafts, Hobart Paper 120, Institute of Economic Affairs, 1993 : a review of why growth rates differ and how to improve economic performance.

(d) **"The Performance of British Manufacturing"** Conservative Research Department, 1993 : corrects some misconceptions and points out positive trends in UK manufacturing.

(e) **"Innovation, Investment and Survival of the UK Economy"**, Royal Academy of Engineering, July 1992. Edited by Ivan Yates: a series of papers exploring Manufacturing and the Economy, Managing Knowledge and Innovations, Financing Innovation and Perspectives for New Policies.

(f) **"Commitment - Implementing the Vision"** Ingersoll Engineers, November 1992 : an investigation of the perceptions of Managing Directors and Senior Managers on the management of change. Supports the hypothesis that successful management of change requires equal emphasis on vision, strategy, implementation, communication and behavioural issues.

(g) **"Made in the UK"** Coopers & Lybrand, Deloitte - November 1992. Survey of British Manufacturing: investigates the competitive pressures on UK industry, the response to these pressures, best practice between sectors, planning for the future.

(h)   **"Innovation, Profits and Investment: the lifeblood of Manufacturing Industry"** - Ronald Garrick (The Fellowship of Engineering, The Engineering Manufacturing Forum Lecture 1992): gives a detailed case history of the Weir Group and is of particular interest on account of its technical and economic details - e.g. on manning and investment.

(i)   **"Industrial Strategy"**, EEF, Autumn 1992 : proposals for recovery and sustained growth. Gives an Action Programme.

(j)   **DTI** - continuing its valuable series of publications on, for instance :

Advanced Manufacturing Research and Development summaries of DTI funded projects, 1993.

Managing in the 90s - The competitive response.

The Carrier Technology Programme - Encouraging the spread of technologies across industry.

Government strategy for IT - The joint framework for information technology.

(k)   "La fin de la Démocratie", Jean-Marie Guehenno, Flammarion; 171 pages, 1993.

12. EXECUTIVE SUMMARY OF 1992 REPORT

"UK Manufacturing - A Survey of Surveys and a Compendium of Remedies"

The question of the continuing decline of manufacturing industry within the UK has been giving serious cause for concern, especially within the electronics industry, for some considerable time. Over the last two years particularly, a number of highly significant reports have been produced by various authoritative bodies which reflect this concern. In order to make some sense of the plethora of information now available and also to provide a source document to assist with the formulation of IEE policy in this area, a report entitled "UK Manufacturing - A Survey of Surveys and a Compendium of Remedies" has been produced by Professor J Levy FEng, under the aegis of the Public Affairs Board.

The intention behind the report was to produce a comprehensive overview of the data available and thus create an independent interpretation of the current situation. Reports utilised include those by the Confederation of British Industry, the House of Lords Select Committee on Science and Technology, the Engineering Employers Federation and the DTI. In the survey a vast body of information has been compressed into an accessible "question and answer" format to get at the truth behind the headlines. The survey seeks answers to the following questions:

1.   Exactly how important is manufacturing to the UK?

2.   How does the UK compare with its international competitors?

3.   How does the UK compare with its competitors on inputs affecting performance?

4.   What proposals have been made in 1990-91 to improve UK manufacturing?

The survey is divided into four sections corresponding to these questions:

1.   **Manufacturing and the economy** - demonstrates unequivocally that services could not replace manufacturing: manufacturing industry accounts for some 62% of UK overseas trade and it would require a tripling in overseas receipts from service industries to supplant it.

2.   **Competitive performance** - the key indicators of:

- output
- productivity
- share of world trade
- import penetration

are the factors which reveal the lower performance levels of the UK as compared with its international competitors. These "indicators" show:

- relatively slow growth in manufacturing output;

- improving productivity, but still below that of West Germany, Japan and the USA;

- a falling share of world trade which needs, at least, to be held at the current level;

- a worsening trade balance.

3. **Inputs affecting performance -**
   Six key inputs are chosen which are under domestic control:

   education and training
   research and development
   innovation
   management style and quality
   size of companies
   finance, investment, Government policy

The survey concludes that:

- the UK work force, apart from the top level, is seriously under qualified;

- the UK has a creditable number of large internationally competitive companies. Twelve of these account for a majority of the national R & D spend and R & D needs to penetrate a wider spectrum of companies;

- the UK is losing ground on patent applications; this indicates a possible loss of innovative capability

- management is improving but further gains are required to raise standards towards those of the best;

- there is a need to create a stronger sector of medium-sized companies;

- there is a need to reassess the balance of interests between shareholders, managers and the work force.

4. **Compendium of Remedies** - this completes the report, and is believed to be a novel approach providing an original and imaginative perspective. It is compiled from a number of significant reports published throughout 1990-91 and assembles a set of potential actions for Government, Industry, Engineering Institutions and other key agencies. These proposals for improvement range from all-embracing socio-economic measures to detailed improvements in manufacturing technology.

13. USA TECHNOLOGY SUPPORT PROGRAM

President Clinton has unveiled a new federal technology policy which he says recognises "that Government can play a key role helping private firms develop and profit from innovations."

His new program, made public February 22nd 1993, calls for focusing on technologies "crucial to today's businesses and a growing economy, such as information and communication, flexible manufacturing and environmental technologies," while reaffirming a commitment to basic science. It also seeks a "closer working partnership" of industry, labour, universities and the federal Government and the States in technology development.

A New Direction

Investing in technology is investing in America's future: a growing economy with more high-skill, high-wage jobs for American workers; a cleaner environment where energy efficiency increases profits and reduces pollution; a stronger, more competitive private sector able to maintain USA leadership in critical world markets; an educational system where every student is challenged; and an inspired scientific and technological research community focused on ensuring not just our national security but our very quality of life.

American technology must move in a new direction to build economic strength and spur economic growth. The traditional federal role in technology development has been limited to support of basic science and mission-oriented research in the Defense Department, NASA and other agencies. This strategy was appropriate for a previous generation but not for today's profound challenges. We cannot rely on the serendipitous application of defence technology to the private sector. We must aim directly at these new challenges and focus our efforts on the new opportunities before us, recognising that Government can play a key role helping private firms develop and profit from innovations.

We must move in a new direction:

- strengthening America's industrial competitiveness and creating jobs;

- creating a business environment when technical innovation can flourish and where investment is attracted to new ideas;

- ensuring the co-ordinated management of technology all across the Government;

- forging a closer working partnership among industry, federal and state Governments, workers and universities;

- redirecting the focus of our national efforts toward technologies crucial to today's businesses and a growing economy, such as information and communication, flexible manufacturing and environmental technologies; and

- reaffirming our commitment to basic science, the foundation on which all technical progress is ultimately built.

For the American People

Our most important measure of success will be our ability to make a difference in the lives of the American people, to harness technology so that it improves the quality of their lives and the economic strength of our nation.

We are moving in a new direction that recognises the critical role technology must play in stimulating and sustaining the long-term economic growth that creates high quality jobs and protects our environment.

We are moving in a new direction to create an educational and training system that challenges American workers to match their skills to the demands of a fast-paced economy and challenges our students to reach for resources beyond their classrooms.

We are moving in a new direction to dramatically improve our ability to transmit complicated information faster and further, to improve our transportation systems, our health care, our research efforts, and even the ability of our military to respond quickly and decisively to any threat to our nation's security.

In these times, technology matters as well to an efficient farm, food processing, and food retailing industry that delivers a variety of low-cost, wholesome foods; to a construction industry that builds high-quality, affordable housing; and to an energy sector that balances energy efficiency with clean, affordable and efficient energy sources.

New Criteria

We will hold ourselves to tough standards and clear vision. The best technology policy unleashes the creative energies of innovators throughout the economy by creating a market that rewards invention and enterprise. We are moving to accelerate the development of civilian technology with new criteria:

- accelerating the development of technologies critical for long-term economic growth but not receiving adequate support from private firms, either because the returns are too distant or because the level of funding required is too great for individual firms to bear;

- encouraging a pattern of business development that will likely result in stable, rewarding jobs for large numbers of workers;

96

- accelerating the development of technologies that could increase productivity while reducing the burden of economic activity on the local, regional, or global environment;

- improving the skills offered by American workers by increasing the productivity and the accessibility of education and training;

- reflecting the real needs of American businesses as demonstrated by their willingness to share the cost of research or participate in the design of initiatives;

- supporting communities or disadvantaged groups in the United States or abroad who have not enjoyed the benefits of technology-based economic growth;

- contributing to USA access to foreign science and technology, enhancing cooperation on global problems or USA successes in technology-related foreign markets.

Reaching our Technology Goals

The challenge we face demands that we set and keep focused on our goals:

- long term economic growth that creates jobs and protects the environment;

- a Government that is more productive and more responsive to the needs of its citizens;

- world leadership in basic science, mathematics and engineering.

We have the means to stimulate innovations that will bring economic growth and help us reach our goals and other important objectives. Foremost is a sound fiscal policy that reduces the federal deficit and lowers interest rates. But that is not always enough. We must also turn to:

- research and experimentation tax credits and other fiscal policies to create an environment conducive to innovation and investment;

- a trade policy that encourages open but fair trade;

- a regulatory policy that encourages innovation and achieves social objectives efficiently;

- education and training programs to ensure continuous learning opportunities for all Americans;

- support for private research and development through research partnerships and other mechanisms to accelerate technologies where market mechanisms do not adequately reflect the nation's return on the investment;

- support for contract R & D centres and manufacturing extensions centres that can give small businesses easy access to technical innovations and know-how;

- support for a national telecommunications infrastructure and other information infrastructures critical for economic expansion;

- Department of Defense and other federal agency purchasing policies designed to foster early markets for innovative products and services that contribute to the national goals;

- strong and sustained support for basic science to protect the source of future innovations;

- international science and technology co-operative projects that enhance USA access to foreign sources of science and technology, contribute to the management of global problems, and provide the basis for marketing USA goods and services;

- dual-Use Defence Department research and development programs;

- national User facilities that make sophisticated research tools, such as synchrotron radiation and neutron beam tools, available to a variety of research organizations.

Managing Technology for Economic Growth

Redirecting America's programs in science and technology will require major changes in the way we manage our efforts. Tight management is essential to ensure that tax, regulatory and other efforts reinforce instead of frustrate our work.

We are making major changes:

Working with Vice President Gore, a reinvigorated Office of Science and Technology Policy will lead in the development of science and technology policy and will use the Federal Co-ordinating Council on Science, Engineering, and Technology, along with other means, to co-ordinate the R & D programs of the federal agencies;

The new National Economic Council will monitor the implementation of new policies and provide a forum for co-ordinating technology policy with the policies of the tax, trade, regulatory, economic development and other economic sectors.

As we move from traditional, mission-oriented R & D to investments designed specifically to strengthen America's industrial competitiveness and create jobs, considerable care must be taken to set priorities. In many cases, it will be essential to require cost-sharing on the part of private partners. In all cases, it will be essential for our Government to work closely with business and labour.

Our initiative in advanced manufacturing, for example, will not be successful without direct input from the private sector about which technical areas are most important. We will conduct a review of laws and regulations, such as the Federal Advisory Committee Act and conflict-of-interest regulations to determine whether changes are needed to increase Government-industry communication and co-operation.

We also will work closely with Congress to prevent "earmarking" of funds for science and technology. Peer review and merit-based competition are critical to the success of any science and technology policy.

Effective management of technology policy also requires an effective partnership between federal and state Governments. The States have pioneered many valuable programs to accelerate technology development and commercialisation. Our efforts should build on these programs.

And every federal technology program, including those of long-standing, will be regularly evaluated against pre-established criteria to determine if they should remain part of a national program. Major changes facing our nation's economy demand a searching re-examination of technology programs, particularly now as we move toward new efforts and a new emphasis in our technology for America's economic growth.

Building America's Economic Strength: New Initiatives

The challenges we face - from our competitors abroad and from our people at home - demand dramatic innovation and bold action that will not just revive our economy now but also ensure our economic growth well into the future. Building America's economic strength through technology demands new initiatives that confront these challenges effectively, efficiently and creatively.

- permanent extension of the research and experimentation tax credit to sustain incentives for the R & D work so essential to new developments;

- investment in a national infrastructure and establishment of a task force working with the private sector to design a national communications policy that will ensure rapid introduction of new communication technology;

- accelerated investment in advanced manufacturing technologies that promote USA industrial competitiveness and that build on, rather than minimise, worker skills;

- re-establishing technological leadership and competitiveness of the USA automobile industry through a major new program to help the industry develop critical new technology that can all but eliminate the environmental hazards of automobile use and operate from domestically produced fuels and facilitate the development of a new generation of automobiles;

- improve technology for education and training by supporting the development and introduction of computer and communications equipment and software that can increase the productivity of learning in formal school settings, a variety of business training facilities and in homes.

- investments in energy-efficient federal buildings to reduce wasteful energy expenses and encourage the adoption of innovative energy-efficient technology.

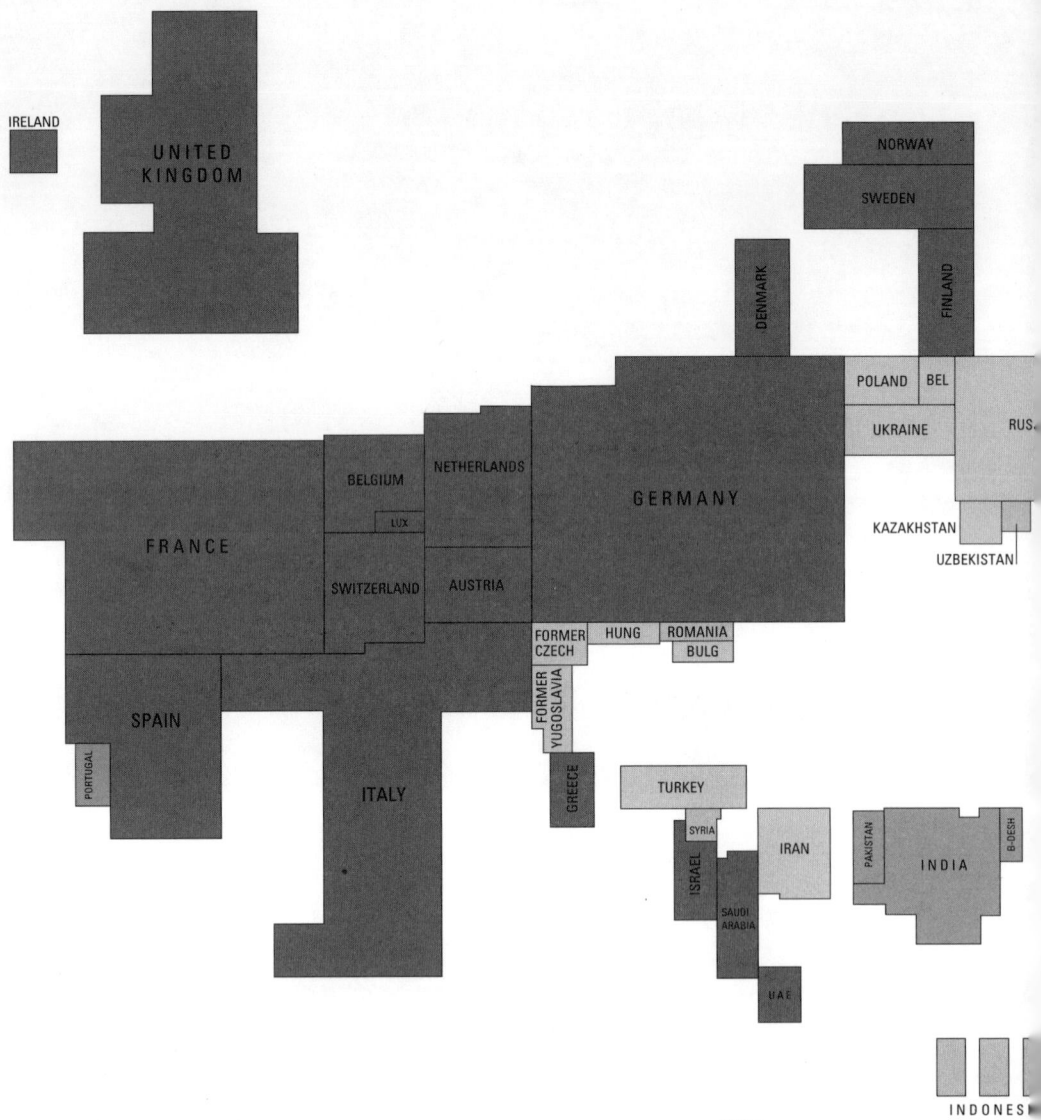

NATIONAL INCOME AND PERSONAL INCOME

SHARES OF WORLD GROSS
PRODUCT *1991* percentages
States with at least 0.1 percent
of total world income

=1%

=0.1%

Source: World Bank

INCOME PER HEAD *1991 US$*

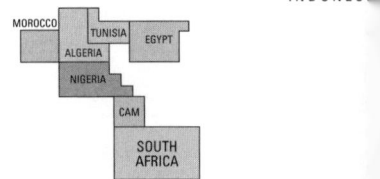

$6,000

$3,500

$1,500

$500

Source: World Bank

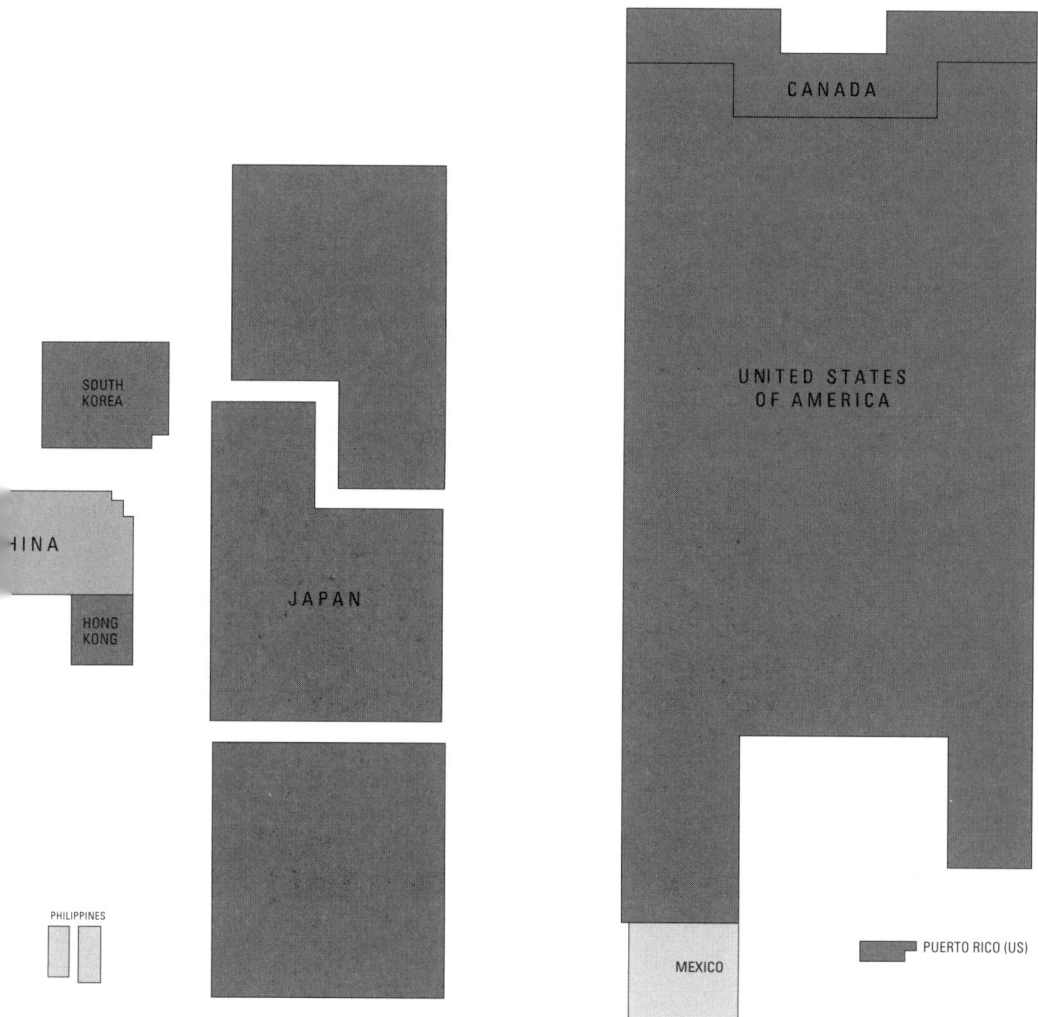

SOUTH
KOREA

HINA

HONG
KONG

JAPAN

PHILIPPINES

CANADA

UNITED STATES
OF AMERICA

PUERTO RICO (US)

MEXICO

COLOMBIA

VENEZUELA

ECUADOR

PERU

BRAZIL

CHILE

ARGENTINA

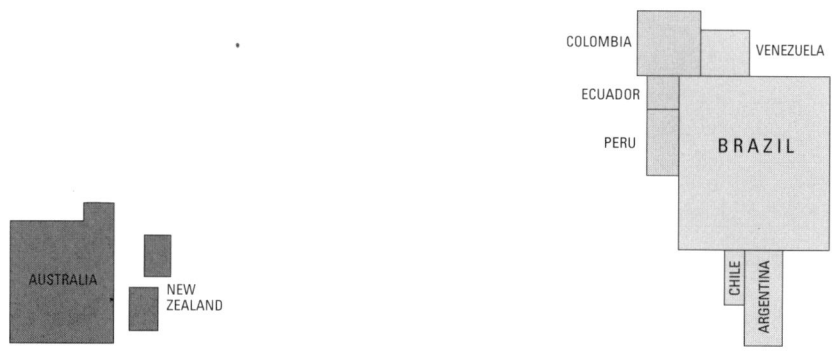

AUSTRALIA

NEW
ZEALAND